EDITION
ESAM

Dr. Elmar P. Selbach

The Millionaire Investor

Better than gold, diamonds or real-estate:

Graphene

How a new disruptive technology is changing the world, and we can become rich with it

Author: Dr. Elmar P. Selbach

www.invest-in-future.info

Published by EDITION ESAM

Cover photo:

(Other pictures in the book are referenced at the end)

ISBN 978-3941769-38-0

How people became rich in the past

We all remember the stories we heard about people who were smart enough to buy stocks of a small company named Microsoft in 1985. At that time, their shares were "penny stocks". By 2000 even a small investor who endured the shares' initial slow growth became rich. Extraordinarily rich.

Who hasn't heard about people later on, and still today restlessly scanning the market for a new "Microsoft", i.e. looking for an investment in a market based on a disruptive technology.

Microsoft itself was a late follower in a technology race that started some decades before.

Bill Gates had been lucky in having acquired a DOS software (DOS stand for Disk Operating System) and by licensing his new tool for IBM.

When that had happened, the Personal Computer era began. Everybody wanted to have a computer on his or her desktop.

Did you imagine such a device in 1980? (P-1)

Apple, Intel, HP and some other companies were part of that game.

The story of Apple is similar to that of Microsoft, when we only look at the value of the shares. Apple today is one of the most valuable companies in the world.

One also could lose a lot of money during that time by betting on the wrong horse.

But let's imagine this scenario: in 1985, you bought shares of 10 different tech companies. An investment of 1.000 Dollar in each of them seemed to be justifiable at that time.

Then you forgot about those stocks and just opened your files again 15 years later.

You smiled as you saw that 9 out of the 10 companies you'd invested in didn't exist anymore or their stock had fallen to a ridiculous amount.

You then smiled more because a company, that did skyrocket, was Microsoft. You lost 9.000 Dollar and won 1.000.000 Dollar

In this book, I want to show that a similar scenario is possible.

One can become rich by investing in the right stock. But not by tricks or cheats, or any miraculous recipes. No.

Simply by observing the dramatic changes in some areas of technology we see ahead of us.

By now we see the dawn of a fascinating new era. The consequences of the changes ahead of us will change our world much more than any event before.

There is a real disruptive technology coming up. This is where intelligent, careful, and patient investors will gain.

I'm not telling you to invest in a specific company. I'm not even recommending stocks.

Since many of the companies I've mentioned in this book are not even listed on the stock market, buying shares would be difficult anyhow.

My goal is to show you the field where the next technological revolution will happen. This book should serve as a first hint where to look when starting investigating new opportunities.

The website related to this book:

http://www.invest-in-future.info

will accompany the companies mentioned below and I'll add new observations as soon as I'll become aware of them.

So this book is dedicated to investors who are careful and don't want to lose money in risky gambling.

This book will be updated regularly with intriguing new developments.

For this edition of the book, 2 chapters were added at the end and information changed within the other chapters, following new insights.

The outline of the book does not follow a linear pattern.

Instead, we are talking first about companies directly involved into Graphene development and deployment.

In a second step we depict companies related to major investment areas such as consumer electronics, automotive or aerospace industry and others.

You'll also find information about recent developments in research and about companies that are trying to establish a Graphene story.

Such the chapters are aligned more or less according to a spiral, showing the exponential development of the market.

First, it seems slow and barely scalable, at the end Graphene will outperform any other market. That's what makes this new market outstanding.

What's the scenario?

When silicon based electronics were introduced at the Bell Telephone Laboratories in 1948, nobody was aware of how this invention would change the world.

Even scientists and visionaries ridiculed the new devices, and it took several years until the value of this technology dawned to leaders of leading companies such as IBM.

From the very beginning in the late '40s until the computer industry started to explode took a full generation – more than 25 years. It was 1975 before most consumers earnestly engaged in computers.

Apple, Microsoft, Intel, HP and all other key players in this field only started to succeed about 25-30 years after the first transistor had entered the world.

Since then, many of these firms have grown from negligible size to some the richest companies in the world. Who didn't dream of having invested just $1.000 in Microsoft in 1985?

Those who did are now counting money on their yacht in St. Tropez or busy improving their golf handicap in Portugal or Australia. Simply said: they are rich.

Gold is a safe harbor for rich people (P-2)

Many people are looking for the next Microsoft to come. What company will emerge from the chaotic ambient noise of stock market newbies?

The venture that will dominate the technology market already exists, or it will be launched within the next few years.

But nobody knows which companies who are just entering the field will be the leading player and make its early investors rich beyond any measure.

The only thing that is clear already today is that it will be an enterprise with a core technology based on barely known materials.

The name for that group of stuff is “**2-D-material**.”

Several chemical compounds and a strange form of the element carbon are listed under that label.

These are e.g. molybdenum disulfide, boron-nitride, phosphorene and a 2-D form of carbon called “**Graphene**”. Yes, Graphene is simply carbon.

We can expect more 2-D material to leave the laboratories of some new and also some old companies.

The main part of this book focuses on Graphene, as this material is known already for its potential to become the super material of the next decades.

So, how will these materials change our world much more than the silicon-based technology did during the last 40-50 years?

This book will describe the stunning discoveries scientists have just made a short while ago, and how during the last three years a complete new industry evolved based on the breathtaking serendipities that emerged from university companies and R&D departments of outstanding players.

It's like somebody finally found the philosopher's stone. Just with the difference that the sorcerers of today's world are real scientists and did not jump out of a Harry Potter novel.

One does not need to know anything about science, technology, or any sophisticated stuff to understand the content of this book.

All you need to have is some sound judgment and savvy business skills, as I'll focus on the companies already developing leading edge products and how they perform in the stock market.

The main focus of this book is on the business aspect of these enterprises and how investors could participate in their success. First and mainly it'll focus on Graphene since everything that is new in the 2-D world is mainly related to this material.

Those who know how to handle Graphene also are capable of producing any other 2-D material.

My goal is to become a shareholder of the "2-D-Microsoft" equivalent of 2020.

Read on if you want to learn more about this unique chance. The bridges are built.

It is worth a Nobel Prize

On December 10, 2010 two physicists entered the stage in the Stockholm City Hall, which was used by the Royal Swedish Academy of Sciences on that day. From the hands of King Carl XVI Gustaf of Sweden they, received the Nobel Prize for physics for that year.

Stockholm City Hall (P-4)

The names of the two scientists were only familiar to a few physicist insiders, and many professors were hit by surprise when the names were published in October of 2010.

The Nobel Prize in Physics 2010 was awarded jointly to Andre Geim and Konstantin Novoselov *"for groundbreaking experiments regarding the two-dimensional material Graphene"* [(1)]

It was mainly then that this phrase became public to the world: “Graphene“. Both Geim and Vovoselov had made their research public in 2004 but only a few companies – mainly from South Korea- initially entered this train.

Today Korean companies are the major vendors for Graphene, closely followed by the USA.

What is it that makes this material so intriguing that some of the largest companies in the world started investing in the emerging technology?

Is Graphene some strange new element brought to us by aliens?

Or did these physicists join forces with witches and magicians in order to make gold from lead, and then ended up with something better?

These physicists found something that will be worth more than any gold mine in the world.

By granting the Nobel Prize to Professor Geim and Professor Novoselov the Nobel committee acknowledged the research both scientists performed on a material they pulled out of a source available to all of us. It is graphite.

This is the stuff you find in any pencil. Yes, it is the small gray stick surrounded by a piece of wood that you use to write your memos.

Before you jump out of your armchair and chase for any pencil in your house, continue reading what the Nobel committee wrote about this invention.

When you've understood that the pathway to success was through a thorough application of the principles of Quantum physics, then you'll also understand that you better leave further efforts to those who are familiar with that kind of theories.

Graphene – the perfect atomic lattice [(2)]

"A thin flake of ordinary carbon, just one atom thick, lies behind this year's Nobel Prize in Physics.

Andre Geim and Konstantin Novoselov have shown that carbon in such a flat form has exceptional properties that originate from the remarkable world of quantum physics.

****

Graphene is a form of carbon. As a material it is completely new – not only the thinnest ever but also the strongest.

As a conductor of electricity, it performs as well as copper.

As a conductor of heat it outperforms all other known materials.

It is almost completely transparent, yet so dense that not even helium, the smallest gas atom, can pass through it.

Carbon, the basis of all known life on earth, has surprised us once again.

Geim and Novoselov extracted the Graphene from a piece of graphite such as is found in ordinary pencils.

Using a regular adhesive tape they managed to obtain a flake of carbon with a thickness of just one atom.

This at a time when many believed it was impossible for such thin crystalline materials to be stable.

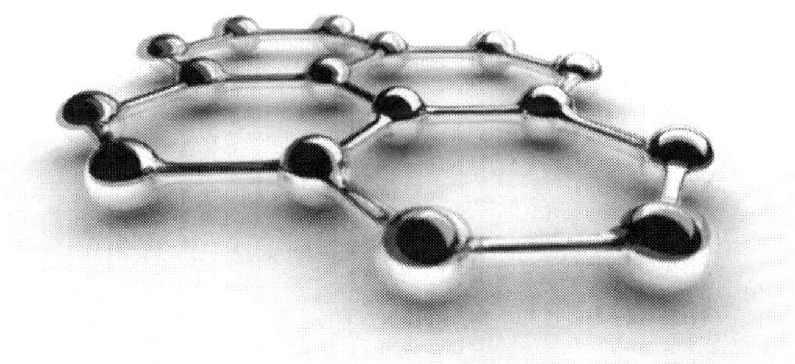

Schematic structure of Graphene (P-3)

However, with Graphene, physicists can now study a new class of two-dimensional materials with unique properties.

Graphene makes experiments possible that give new twists to the phenomena in quantum physics.

Also a vast variety of practical applications now appear

possible, including the creation of new materials and the manufacture of innovative electronics.

Graphene transistors are predicted to be substantially faster than today's silicon transistors and result in more efficient computers.

Since it is practically transparent and a good conductor, Graphene is suitable for producing transparent touch screens, light panels, and maybe even solar cells.

When mixed into plastics, Graphene can turn them into conductors of electricity while making them more heat resistant and mechanically robust.

This resilience can be utilized in new super strong materials, which are also thin, elastic and lightweight.

In the future, satellites, airplanes, and cars could be manufactured out of the new composite materials.

This year's Laureates have been working together for a long time now.

Konstantin Novoselov, 36, first worked with Andre Geim, 51, as a PhD-student in the Netherlands.

He subsequently followed Geim to the United Kingdom.

Both of them originally studied and began their careers as physicists in Russia.

Now they are both professors at the University of Manchester.

Playfulness is one of their hallmarks, one always learns something in the process, and who knows, you may even hit the jackpot.

Like now when they, with Graphene, write themselves into the annals of science."

(Cited from nobelprize.org [2])

The most intriguing fact is that Graphene is just simply carbon.

This black material you found in your grill after last Saturday's barbecue.

Or that stuff you buy in a jewelry store in the form of a diamond.

Or that you write with it when you use a pencil.

So what is it that makes Graphene such a wonder material?

Basically, it relates to its 2-dimensional appearance. Diamond and graphite are simple bulk material, i.e., it appears in 3-dimensions.

This is apparent in diamonds' excellent heat, transportation capabilities, hardness and transparency.

But, diamonds are missing all of the other features of Graphene.

In addition, Graphene is cheap compared to shiny jewelry. Any factory with a sophisticated 2-D manufacturing process can produce it, even though in small portions (which is about to change, soon).

Got it?

Any factory with a sophisticated 2-D manufacturing process can produce it. This is essential.

What company can show us such a sophisticated process?

There are only a few and again, most of them are located in Asia, with some in Europe and the USA.

A 2-D process is still far away from being a simple method. This is high-tech at the finest.

What Google and Eric Schmidt do not know – yet

Google CEO Eric Schmidt recently spoke in an interview about disrupting technologies.

He stated that technologies such as cloud-computing and digital biology were the most compelling technologies that would revolutionize the world.

In addition, he mentioned 3-D printing.

He is right bringing up all of these areas. But he doesn't realize that the development that has just started in the area of 2-D technology will dramatically influence the changes we'll see.

Especially 3-D printing technology will benefit from Nano-material science and within that realm, again, particularly from Graphene.

In one of the chapters below I'll talk about companies that already created first devices by combining 3-D printing technology with 2-D material science, using Graphene and not simple plastic.

Companies who have invested in hard-disk based storage farms or classical semiconductor devices will have to undergo a dramatic refocusing on new development areas – all of them triggered by 2-D technology.

Among scientists, there is no doubt that Graphene and related 2-D materials, such as Boron-Nitride, Molybdenium-Sulfide and others will revolutionize the world in terms of information and communication, energy storage, energy generation, and many other areas of our society.

This revolution will start with the material science, the information and communication area, immediately followed by developments related to energy harvesting and storage.

Digital biology and the integration of biological material such as DNA with the 2-D material will allow computerized cells.

Later in this century, we'll see devices that we'd never thought could exist.

Electronics will become wearable, such as foldable mobile phones or computers one can fold like a sheet of paper.

Military robots built with Graphene wings will fly autonomously for weeks.

Similar flight robots could serve as routers for wireless wide area networks, making neighborhoods and towns independent from carriers.

Even small computers will have features known to the most expensive devices today: 3-D displays with brilliant graphics and a computing speed of today's supercomputers.

Realistic 3-D technology through Graphene displays (P-7)

If you think that this is just blarney of a bragging gasbag, you should read the reports written by the scientific committee for the Graphene flagship project recently established by the European Commission.

This organization announced in 2012 the spending of 1 billion Euro for only one research topic – Graphene based technologies.

Yes, the number above is correct: 1 billion Euro. At the time I was writing this, that amount is worth 1,320 Billion US$.

Ever seen US$1, 000,000,000 in real money? (P-6)

Has anybody ever heard of a similar plan of the government in the United States?

Only DARPA has granted some funds for research in Graphene, so far.

Why did a state agency in Europe put so much money on the table?

In addition, large companies such as IBM or Samsung are investing billions of dollars just in 2-D scientific projects.

All this is possible because Graphene and other 2-D material will bring a new dimension to technology.

Future material and devices will not just be gradually smarter, faster, thinner, safer, stronger, and more intelligent.

They will exponentially outperform today's technology.

A similar disruptive development took place in the 1970th, when companies like IBM, Texas Instruments, DEC, and many others were able to integrate semiconductor devices into small chips, such as microprocessors.

Since then, the speed and performance of these products have grown exponentially.

Devices are by factors of many millions faster, thinner, and more intelligent than anything what was available before 1975.

The next leap based on 2-D material will even be bigger.

The European Commission calls Graphene research an example of an emerging "translational technology" [4].

That is a common phrase for scientific discoveries, which will fast be transferred to technological and market applications.

Based on that research, Graphene will be able to improve and speed up existing silicon semiconductor devices, such as memory chips and microprocessors.

A short time after these first steps, entirely new developments and applications will come up.

If you want to get a glimpse on just one of the first steps, see the YouTube video published by Nokia in January 2008 about their vision. [5]

At that time, they'd no clue about Graphene (they were only mentioning nano products) – but they had this vision.

> *In 2011 Nokia stumbled into Graphene.* [6] *And in 2013 stumbled out again, but this is another story.*

Now with the availability of 2-D material their vision is close to becoming a product, but will be produced by more intelligent companies.

Graphene not only offers stunning electrical properties, but also will become a killer application in optical technologies.

Future devices will operate at breathtaking speed and offer brilliant graphics with transparent and flexible systems.

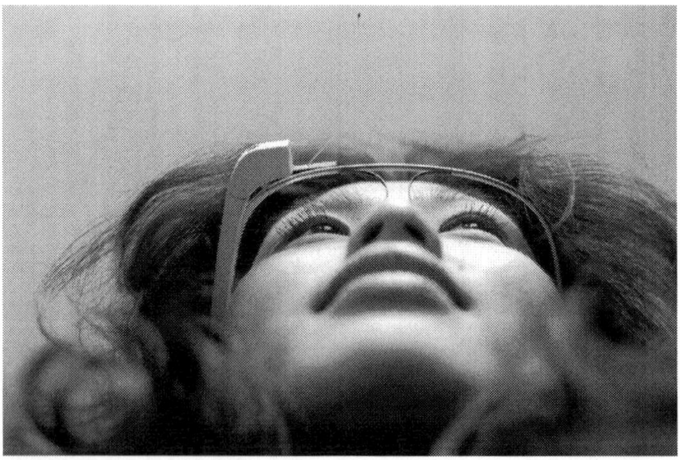

Google Glass – eventually made of Graphene? (P5)

Just think of a foldable supercomputer on your wrist, which can send information to nearly undestroyable glasses – all made of Graphene.

Google glasses are coming into your mind? You are already on the right track.

Graphene is not only another new electronic material.

It offers more features than any other material and together with Boron-Nitride, Molybdenium-Sulfide and other 2-D materials, it will revolutionize the way we safe energy, produce power and get rid of poisonous materials such as waste from chemical or biological plants.

A young scientist in 2014 proposed a method of cleaning pollutant areas and material using a mixture of Graphene and Titanium-Oxide.

Already today, we can see Graphene-based solar panels emerging in the near future, helping to increase the amount of energy produced on roofs and in solar-farms by a factor of about 100.

This new technology will also be much cheaper than silicon based solar cells, and because of their flexibility, can be mounted anywhere onto a building.

We will not need more oil, gas, and power lines, because the energy harvesting concepts of the future will turn any building into a self-reliant power plant.

What else is up?

Graphene can work as a catalyst, breaking poisonous chemical compounds into harmless pieces of water and carbon-dioxide.

It could probably also be used as storage for liquid hydrogen.

Today's hydrogen storage systems are extremely expensive and also dangerous.

Embedded into a Graphene matrix, hydrogen storage will both be safe and cheap.

Fuel cells based on Graphene will change cars, and that is for sure already.

Actual research shows that fuel cells using Graphene membranes can improve energy capacity and durability by a huge amount.

Think of an electrical vehicle powered by liquid hydrogen. This hydrogen will produce electricity in a fuel cell that will drive 4 small electrical motors, one in every wheel.

These cars will be ultra-light compared to the vehicles available now. They will be safe, and have energy efficiency of more than 80%.

Today your gas driven car –even if you are driving a hybrid – will not reach a level of efficiency larger than 35%.

Your car is wasting more than 3-thirds of the gallons it swallows.

With Graphene based super fuel cells, no more of your money will vanish through the exhaust pipe.

Future cars will not need an exhaust pipe anymore. One is only using that amount of energy needed to move and heat or chill that vehicle.

Using Graphene in a car will enable it to have ten-times as much storage capacity in its batteries, and also speed up the charging process by the same factor.

Autos will weigh 50% less than metal-based cars, if its chassis, and drives, and motors are made of Graphene.

At the same time the cars will become much safer because of Graphene's extraordinary strengths.

By integrating fuel-cell technology with – let's say an energy harvesting system in your house – you never will use more energy than the sun is delivering.

Also, the wheels of future cars will be made of a Graphene derivate.

This 2-D material is much stronger and mechanically resistant than any other stuff available in the world and at the same time so cheap that everybody can afford it.

Usage of Graphene for tires will make them lighter, shortening the stopping distance and enable longer usage cycles.

Even streets could be built of Graphene. Using the electrical features, one could make such streets to be solar cells at the same time.

Similar progress for airplanes is not a dream, but a vision that has made its way into the engineering departments of manufacturers such as Airbus, Boeing or China Aerospace.

Think of planes that are much stronger, and at the same time much lighter than today, driven by propellers powered by Graphene batteries.

Graphene's main application will be in electronics (be it communication technology, Internet of Things (IoT), computer devices or sensors, 3D printing and in renewable energy.

It will replace silicon and germanium as well as compound semiconductors (such as Gallium-Arsenide, Indium-Telluride and other exotic products) in computers and telecom devices.

Solar cells will catch 20-times the amount of energy today's silicon based system can harvest based on combined Graphene / Molybdenium-Sulfide technology.

The energy could be stored locally and any house – even in Alaska – could become independent in terms of electrical power.

Solar panels will cost a fraction of silicon based systems and also be much more durable.

More applications are on the horizon and as I write this, there are many scientists all over the world already thinking about new products and use-cases for 2-D material.

Graphene and Boron-Nitride could change the way doctors heal wounds and repair broken bones.

Medical instruments will make use of the tremendous features of Graphene.

Biologists will dramatically improve the way they are decoding and synthesizing DNA and proteins.

And these are only a few applications.

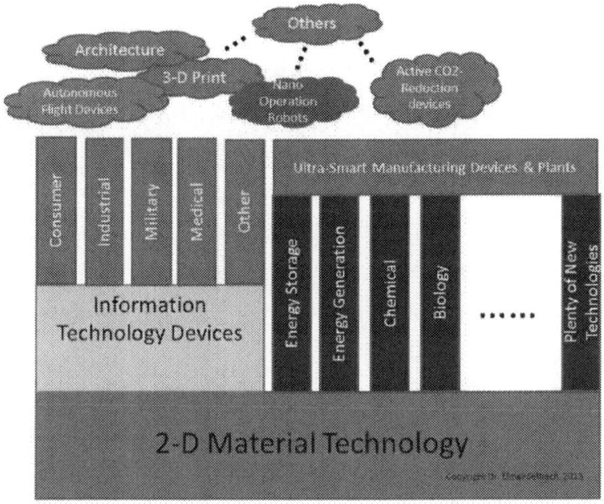

Potential application areas for 2-D material [(P-8)]

By 2014, we have seen up to 100 startup companies entering the market with inventions based on Graphene.

The key players such as Sony, BASF, IBM or Samsung, will not be able to finance every possible application their researchers are going to find.

So a lot of spin-offs will emerge from those labs where aspiring engineers and scientists will flood the market with their fantastic ideas.

> *(In the 1980's I was part of a research team at a university that worked on a brand new technology called Chemical Vapor Deposition (CVD) of a special kind of semiconductor devices.*
>
> *The company emerging from that team later took the name Aixtron. Large companies at that time never came close to what had been invented by our team.)*

Today Aixtron is one of the key players in the Graphene market itself.

By the way: CVD will play a crucial role in synthesizing Graphene based devices.

At least one of the entrepreneurs of these companies will run a company that will become the 21st century equivalent of Microsoft.

This is a company that will outperform any other competitor and has stock prices that will grow exponentially.

The name of this company is still unknown, but with a thorough research and patient investigation of the key players one will soon get a glimpse of the market development.

Apple and Google should be among the winners.

The graphics below show some potential and foreseeable applications over the next 25 years.

As any of such predictions look somewhat ridiculous when one looks back, it is already clear that the next generation will laugh about these graphs.

Reality by then will outperform our predictions by far.

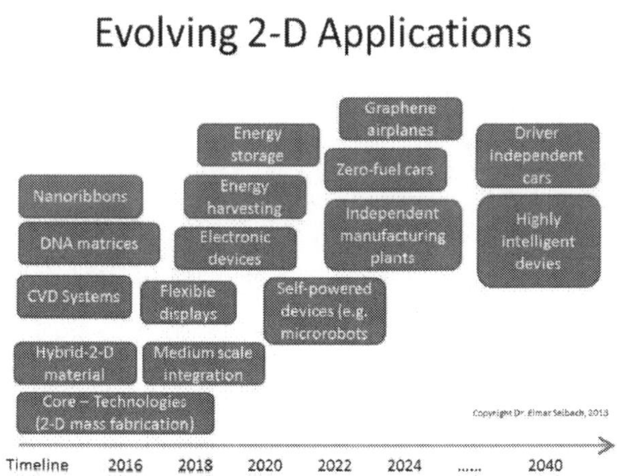

How 2-D applications could evolve over the next decades (P-9)

How disruptive developments change the markets

Famous Quotes:

"Everything that can be invented has been invented." *Charles H. Duell, Commissioner, U.S. patent office, 1899* (7)

"I have traveled the length and breadth of this country and talked with the best people, and I can assure you that data processing is a fad that won't last out the year." - *The editor in charge of business books for Prentice Hall, 1957* (8)

"But what... is it good for?" *-Engineer at the Advanced Computing Systems Division of IBM, 1968, commenting on the microchip.* *(9)*

"There is no reason anyone would want a computer in their home." -- *Ken Olson, president, chairman and founder of Digital Equipment Corp., 1977* *(10)*

"Radio has no future. Heavier-than-air flying machines are impossible. X-rays will prove to be a hoax." -- *William Thomson, Lord Kelvin, British scientist, 1899.* *(11)*

And so on....

During the last centuries the world has seen many disruptive technologies such as the personal computer, the Internet, self-driving carriages (aka automobiles) and electrical devices of any kind.

During their invention phase, many (even highly educated) people were not able to make any use of the related inventions.

These people publicly neglected the right to exist for the inventions mentioned at the beginning of this chapter.

The best ever given reply to such a question was given by Michael Faraday, when the British Chancellor of the Exchequer (minister of finance), William Gladstone, questioned the practical value of electricity: "One day sir, you may tax it." (12)

Here we are. The goal of this book is to find companies to invest in and then become rich from stocks.

The treasury secretary or ministers of finance then may come to take their share.

If that will be the only difficulty we are going to encounter, we should then call ourselves lucky.

But there will be many pitfalls before a mature, stable 2-D business will dominate the stock markets.

What is already foreseeable today is the size of the market.

According to the German chemistry giant BASF, just the Graphene market will reach a size of 1.55 billion Dollars by 2015, which is considerably small if one compares it to the silicon market.

Already by 2020, the Graphene market will have grown to about 8 billion Dollars.

This is an increase by the factor of 6 within 5 years. This momentum will continue over the next decades.

This forecast just addresses the core market, such production, and a few related technologies such as CVD.

If Graphene is really an upcoming disruptive technology, it must enable many applications and additional technologies. In the last chapters I've outlined a few.

Graphene is the only known technology today, which shows the potential to become an overwhelming success story in the next decades. This is also, because it can perfectly be combined with technologies like 3-D printing.

Its properties are stunning and are unknown in any other material so far.

Graphene is thin. Thinner than anything else in the world. In fact is only one atom in a layer with other atoms that lie flat on a surface.

That is the reason it is called a "2-D" material. 2-D here is the mathematical term for Graphene being so thin that it only exists as a flat film.

Bulk material such as iron or diamonds, apples, trees and anything else are 3-D things because they feature 3 dimensions: width, height and length.

Our hero material here is made 100% of carbon in the same form it exists in graphite.

That is something everybody has in his or her household because it is the core of any pencil.

Much of the technology expected to boom with Graphene is related to producing Graphene out of graphite, i.e., bringing this bulk material in an absolutely flat form.

Carbon is also the element diamonds are made of.

Graphene is very hard. Actually, Graphene is much harder than diamond and at the same time it is much more elastic than anything else.

When electrical currents are applied to it, they flow with a breathtaking speed through the thin layer.

Graphene holds any records in speed of electrons, “virtual” mass (mobility) and capabilities of switching currents on and off.

The latter is the basis for building transistors. Today, more than 99% of all electronic devices in computers or embedded systems is based on transistors.

Those are connected on microchips such as DRAM (memories) or microprocessors to large computing machines. Most of these devices today use silicon as the base material.

Graphene based transistors show such superior capabilities over silicon that it is clear that the future of electronics is with Graphene. It is capable of switching 100 times faster than silicon.

(There might be an exception to this inevitable upheaval. As of 2012 another material called “Silicene” sprang on stage. Both Graphene and Silicen together might be able to find their way. I’ll talk about it at the end of this book).

Silicon ingots: those will become obsolete upon the arrival of Graphene [P-12]

These features also will make Graphene the dominate solar cells.

Solar farms could be established on every roof and would provide multiples of the energy present day's technology is capable of delivering.

Future houses and towns could become completely self-sustaining and independent form power providers.

Graphene also is optically active, chemically absolutely inert and does not allow any gas to permeate it. Optical activity is the basis for any modern display technology.

Being chemically inert means that one can use this material in high-end biological and chemical synthesis processes.

No other material has this capability. Membranes based on Graphene will enable chemical and physico-chemical reactors (such as fuel cells) which will revolutionize chemistry, energy transport and energy storage.

Intelligent engineers and scientists will find more applications than we can think about, since for the first time in history a material with such a multitude of special properties has been found.

Eventually copper will not be used anymore for wires, batteries, or electrical interconnects.

Graphene based batteries will revolutionize the way we store energy. [(P-10)]

Graphene does it better, is not being destroyed in airy or wet environments and is much, much cheaper than that metal.

The electrical current density of Graphene is more than a million times greater than that of copper.

Petrol will become useless, because Graphene based energy farms and energy storage devices enable hydrogen – which is made of water – to become the fuel of the future.

Storage of that gas will be safe and fuel-cells will provide more power than any gas-driven motor is capable of delivering. (12)

From what is said so far, what industries will benefit firsthand from Graphene?

The semiconductor technology, the telecommunication provider (broadband speed of 10.000 gigabit and above will be possible), the chemistry industry, and automotive and aerospace industries will become winners in the upcoming race.

Consumer goods such as mobile phones, gadgets and everything else one has in the house such as dishwashers or fridges will look completely different from today.

The DVD player and the TV of the future will be integrated into your wall as a sheet not thicker than a piece of paper. Other such gadgets will be foldable.

Amazon's Kindle or Nook e-book readers will be nothing more than a light sheet.

An iPad might weight not more than a small pocket book or brochure and being flexible at the same time.

The road to this new world is long.

Don't estimate this vision to become well established before 2025.

The only thing we know today is that everything outlined above is technologically possible.

Since 2004 the world has the material which is required to fulfill our dreams.

Before selling your stocks of mining companies and power and utility companies, wait some years.

First the technological problems in Graphene production and manufacturing of its deriving applications need to be solved.

But don't wait too long. 2-D technologies will change everything we know. That is an already given.

The only open questions are how long it will take and what application areas will be revolutionized first.

So before we investigate the potential "Microsoft-equivalents" of 2030 let's have a look on the upcoming application areas.

Below is a listing of technologies, applications, and companies that become interesting under the perspective that is outlined so far.

The list is far from being complete, and some of the companies and institutions mentioned above haven't entered the stock market yet.

Please observe that I'm not giving any recommendation for buying or selling stocks of those companies.

You should take the time to start your own investigation process and follow these entrepreneurial high-potentials, using your own wisdom.

As far as I'm an owner of stocks of companies I talk about, I've mentioned it below.

During the outline of this book I'll give an overview you what company stocks I'll buy when the time has come.

Many of the companies I'm observing have not yet entered the stock market. The firms I'm mentioning are those, which will be on my investment radar.

Some businesses are described in the next chapter. For the majority of the companies I'll update a section of the website: (http://www. invest-in-future.info)

(At the end of the book you'll find also the web links to the companies I mention in this book)

2014: Companies earning money with Graphene

Now is the time for engineers in base technology firms to get Graphene up and running. Those companies are either in the market for a long time (like IBM) or have just recently opened their doors.

Most of the Graphene business today focuses on mining companies.

The major portion of Graphene in the market is taken from graphite, and this core-product is dug out of the earth with conventional methods.

The mining companies are mainly in Canada, India, and China.

Graphite is refined in a chemical and physical process and then sold to those companies that are developing applications.

Among the miners and refiners are: [14]

Asbury Carbons

Focus Graphite

Grafoid

Lomiko Metals

Northern Graphite

Syrah Resources

Zenyatta Ventures

Graftech International (GTI)

Those companies are mentioned on many websites and are recommended by many stockbrokers as lucrative investment opportunities in the emerging market for the "wonder-material" Graphene.

Here is my opinion about investing in these companies:

I will NOT buy any stock of those and similar firms.

Why not?

There are two reasons for that:

The first has to do with the unique properties of Graphene and other 2-D material.

Within 10 years from now Graphene could come from large scale CVD systems, which pick carbon from organic and inorganic sources available everywhere, including CO_2 from the air (the latter technology is already being tested in plants in Germany and Japan. Excess power from wind turbine farms is used to extract CO_2 from the air and to refine it into carbon).

Additionally, bio-ethanol should become a high-quality source for Graphene. Vendors are already in the market. (See below)

Second: any mining company might go (and probably will go) through the same stock cycle, as other mining companies are known for.

That means constant ups and downs, erratic breaks of the prices, triggering by crazy fund-managers and so on.

A similar hype we've seen around rare-earth metals during the past years.

Driven by hysteric rumors, the press published reports suggesting that the industry in the western world would go to hell by lack of rare-earth metals. But nothing happened.

Only the backbiters of the false information finally laughed about the stupidity of those people buying rare-earth investments for ridiculous high prices.

Some risk-adverse gamblers might try their luck within that market. I definitely won't spend even one cent for those producers.

There are many producers of bulk graphite who just made slight changes to their company presentation, bragging to be "Graphene" suppliers.

Graphene business is a high tech business and just dressing the shop window into other colors doesn't turn a barber shop into a university hospital.

On top of that China announced in 2013 that it might be capable of producing tons of Graphene in large-scale production facilities within 5 years.

That is another reason why I would always go "short" for mining and refining companies.

I'm looking for stocks taking off from the beginning of their business – and those having a bright future.

Those, that can't be squeezed by any broker or hedge fund.

I have reasons enough to doubt that mining companies will ever belong to the class of stocks I prefer, as I'll outline below:

> *A short digression: At the beginning of the 20th century, most of the fertilizers farmers used were imported from Chile. The sodium-nitrate mines in Chile also delivered the base material for the allied bombs and gunpowder.*
>
> *Their enemy Germany was cut off from any nitrous source by 1914.*
>
> *By then the Germans switched to extracting nitrogen – which is the base of all fertilizers and also the main ingredient in gunpowder – simply from the air, using a process invented by a scientist called Fritz Haber just some years before.*
>
> *Chile's saltpeter mines became nearly obsolete afterwards.*
>
> *Today the majority of fertilizers and any nitrogen-based chemical are produced from nitrogen pumped out of the surrounding air.*

Another example:

In the past years we saw a hysterical race for rare-earth metals, which are mainly available in Chinese mines.

Other than the naming of metal like Erbium, Cer and so on indicates, rare-earth metals are anything but rare.

They can be found everywhere. The only challenge is to extract them from the environment they can be found in.

Rare-earth metals are abundantly available in seawater.

By using excess power from wind turbines in the North Sea, countries like Norway, UK, Holland or Germany are able to extract that material from the ocean.

This process has just started to be investigated.

But knowing how easy it is to extract salt from seawater and in a second step to dissolute the salt into metal and halogen, it is obvious that mining doesn't have much of a bright future.

Any other metal can be found in seawater as well.

One day there will be more than enough power from wind turbines and solar fields available. That might be the end of any mining company, as we know them.

Mining might become obsolete one day – Graphene can replace most metals – even rare-earth metals (P-11)

There are other potential candidates to invest in today when it comes to core-products.

They are closer to technology, have an established product line already and are driven by visionary managers.

Lomiko, one of the producers of basic Graphene, which I mentioned above, already understood that the value in the 2-D material market would be created by companies building sophisticated applications using Graphene.

In 2014, Lomiko bought more than 4 million shares of Graphene 3D Labs Inc. ("Graphene 3D") (TSX VENTURE: GGG), sending the penny stock of that company to new heights.

This might differentiate Lomiko eventually from the mining companies and therefore they are remaining on my radar screen.

There are some question marks related to this investment, anyhow. I'll talk about those when presenting Graphene 3D Labs Inc.

My final conclusion of that chapter is that one should look for companies that take Graphene as a raw material and develop high sophisticated applications and products from that.

Such as the big wins in the silicon area didn't come from sand (the raw material for silicon) digging firms, but from high-tech firms such as IBM or Microsoft.

Investment candidates might be these companies:

Small companies to observe: (14)

Aixtron

Applied Graphene Materials PLC

Graphene 3D Labs incorporated

Vorbeck

Graphene Nano Chem

XG Sciences

CVD Equipment Corporation

GRAnPH Nanotech

Crayo Nano

Angstron

AMO GmbH

SouthWest NanoTechnologies Inc

Nantero Inc.

Not every of these companies is market ready yet.

As of today, most are not tradable on the stock market and those where one could acquire shares, are too often still risky investments.

But it might be worth keeping an eye on most of them, (e.g. **AMO or Graphene 3D Labs**) since they might become key players in their respective niche, when they enter the stock market, or when a big player swallows the company.

Aixtron

This is a company I knew well when they started into business.

I was a part of the research team that worked in a university laboratory in 1983.

Aixtron is a key player in CVD technology. CVD stands for chemical vapor deposition and is a key technology when ultra-clean and perfect Graphene layers are required.

Other than the companies mentioned below, Aixtron is publicly traded since 1997.

The CVD market is very much related to the semiconductor market and shows the same hefty up and down of those stocks.

As of autumn 2014 the shares are very low, as the CVD market in total is suffering from low demand of their semiconductor markets.

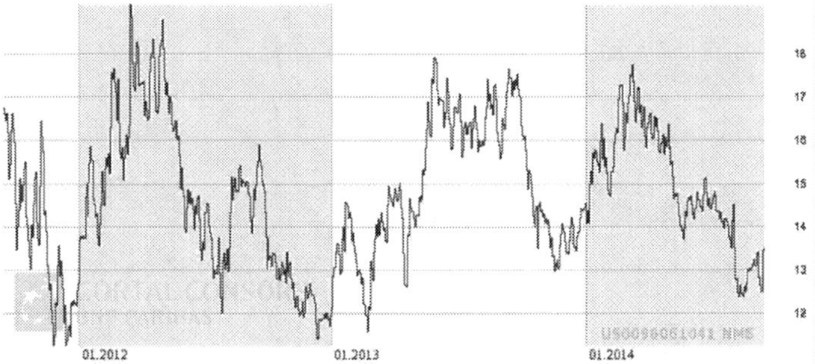

3-years performance of Aixtron stocks

Aixtron is listed on the NASDAQ (AIXG).

They have a very strong R&D background. (By writing “very strong” I mean very strong. No bragging.)

Still being connected to the Technical University in Aachen (Aix La Chappelle in French), they are able to adapt their manufacturing capabilities to the compelling demands of the Graphene industry.

Analysts predict a cash flow margin of 4.5% in 2015. (59)

Photo: CVD reaction chamber of one of Aixtrons' key products

(By permission of Aixtron SE)

Applied Graphene Materials PLC

That company tells us on their website, that they've "developed a proprietary "bottom up" process for the production of high specification Graphene. Applied Graphene Materials owns the intellectual property and know-how behind this process." *(56)*

My focus of this company is based on their know-how in creating basic material from Graphene for 3-D printing.

Their stock history looks somewhat sound:

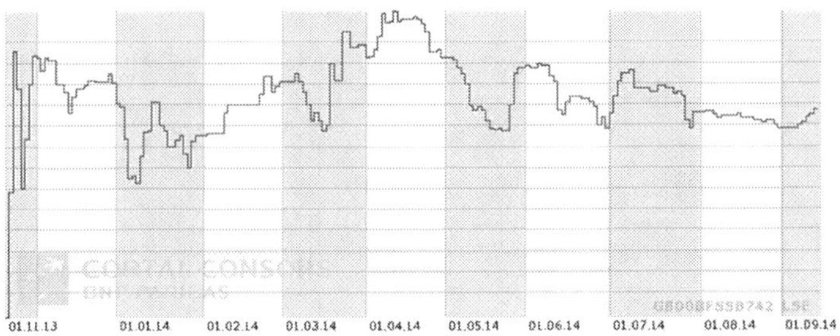

3-year stock performance of Applied Graphene Labs PLC

The stock graphic not showing many ups and downs, this company eventually might slowly gain momentum, based on new inventions from their R&D labs.

Offering a highly sophisticated method for refining Graphene, which is not taken from graphite, but organic sources (a big differentiator...) the stocks might rise along with the Graphene market maturing over time.

Three financial investors and an IPO firm own the majority of the stocks from the UK. About one third of the stocks is freely traded.

The market capitalization is about 70 million British Pounds, which is about 114 million US$.

In 2013 the company reported a slight loss, but insignificant as it was, the stocks remained stable.

Graphene 3D Labs Inc.

Calverton, New York, based Graphene 3D Lab recently started as a spinoff from Graphene Laboratories, Inc. This company claims to be a "worldwide leader in manufacturing and retailing Graphene and advanced materials". (60)

I added Graphene 3D Labs into this book because they seem to be capable of combining Graphene technology with the emerging 3-D printing market.

At least that is what they are claiming to be able to do by 2015. But there is some grain of salt added to the full story.

The stock performance over the last 3 years is a show-stopper – at the first look:

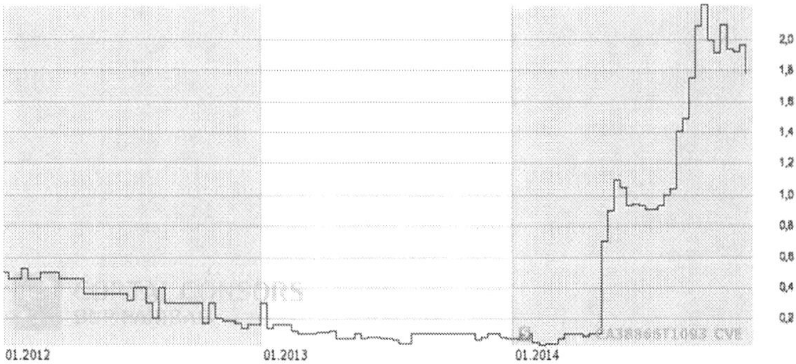

3 Years stock performance of Graphene 3D Labs Inc.

Something happened in February 2014.

This was when previously mentioned Lomiko acquired the majority of shares of Graphene 3D Labs.

They bought it for less than 10 cents per share, when my calculations are correct.

An intelligent decision by the bosses of Lomiko, as the company they invested in claims to have a unique technology for creating 3-D products with Graphene.

It seems to have paid already, as the shares rose to more than $2.

These 3-D products might be batteries, consumer products or parts for robotics applications, among others.

From my perspective, Graphene 3D Labs are one of the most interesting startups one should consider investing in, if the story behind the company finally keeps all its promises.

According to their website, Graphene 3D Labs, mentions the following customers: NASA, Ford Motor Co., Honda, GE, Apple, Xerox, Samsung, Bosch, LG, Harvard University, IBM, MIT, Caltech, Yale, Princeton and Stanford Universities. [60]

Along with the management team's vision and the person of the CFO there is no indication that something might be fishy with that company.

This was it for the positive part.

As of writing this, the investment in Graphene 3D Labs shows a lot of risks, which should prevent any careful investor from buying stocks at this time.

Trading at more than $2 per share, with a market cap of more than $40 million, Graphene 3D Labs looks like a lucrative investment.

Graphene 3D Lab's financial statements dated from July 2014.

The company claimed to have assets of $1.3 million, losses of $1.6 million and no revenues. (See Reuters.com for details).

According to their website, the company Graphene 3D Lab has filed several applications for patents, the latest on September 2, 2014.

What puzzles me is that I could not find one of the patents filed in the past year in the database of the US patent office. (http://www.uspto.gov/patents/process/search/)

So the financial facts are not very intriguing (the technological story, on the other hand, is very compelling...).

- Stocks rose from $0.02 to about $2.40 within the last year.
- Patents are pending.
- The owners of the company, the CEO and the COO, seem to be a married couple, which might lead to conflicts in the company strategy
- No revenues expected before 2015
- The R&D team mainly seems to consist of the CEO and COO (it's open how the backlink to the university they are coming from works)

Simply said: despite the very interesting product announcement and great opportunity depicted in the company slides, I'll wait before buying stocks until the first products from Graphene 3D Lab have successfully entered the market.

As a value investor with a slight inclination to risky investments, this is still a story too difficult to swallow.

May be I'm too late when the stocks finally explode. But as said before, this investment looks very risky to me.

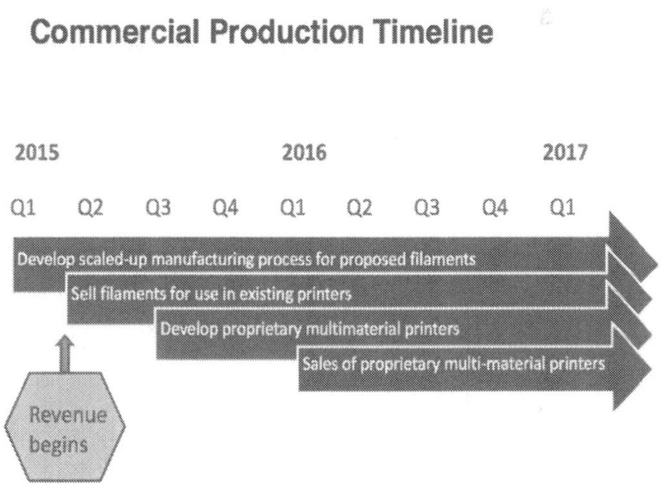

Timeline 3DLabs revenue plan (source: Graphene 3D Lab investor presentation, by permission)

According to the plan of that company, revenue will start by mid of 2015.

This will also be the time I'll set an action to check the shares again.

Then the pending patents should also appear in the database of the US patent office.

If that'll happen as outlined above, Graphene 3D Lab definitely will be a prime target for investments, as by then they might be the only company capable of combing 3D printing with 2D nanotechnology.

Vorbeck

This company from Maryland, USA, uses Graphene technology to develop conductive compounds for applications in the electronics industry.

It was founded in 2006 and licenses its technology from the University of Princeton.

Vorbeck has a 40 ton per year Graphene production capability as of October 2012. *(61)* They produce conductive ink (Vor-ink ™) and coatings (Vor-coat ™).

Potential customers in that area are RFID vendors, the packaging industry, sensor companies (both passive and electronically controlled), wearable technology firms (which is interesting for the military industry), packaging, electrical storage (batteries and super capacitors) and medical organizations.

Think about electronic products as thin as a sheet of paper and also flexible, without being destroyable. That is what one could achieve with Vorbeck's Graphene solutions.

From my perspective, a product like their conductive ink, for instance, could be a core product in new consumer gadgets like future versions of the Apple watch or new versions of eBook readers such as Nook from B&N or Amazon's Kindle.

In an interview with the BBC the president of Vorbeck, John Lettow, demonstrated a sheet of Graphene prepared for use in electronic devices.

In front of the camera, he folded and crushed the device, which sounded like ripping a piece of paper.

But this sheet remained totally stable and unharmed. According to Lettow, you could even put it in a washing machine or iron it without destroying.

The main commercial application could become batteries, anyhow.

Using Graphene, the charging and the discharging behavior of batteries will completely change.

Instead of hours, it'll take only minutes to re-charge a battery and maintain the power for a much longer time.

And for all of the new products one would only need a manufacturing process that is not more complicated than printing a newspaper.

Combining Vorbeck's technology with that of companies in the 3-D printing market, such as Graphene 3D Labs

Inc., will change the consumer market in a way never would have thought to be possible.

Vorbeck Materials Corp. completed a financing on December 9, 2011, with a total value of $10 million. Main investors at that time were Black Powder, LLC and Fairbridge Venture Partners, and 15 additional investors.

It is not known to me when this company ever will go public. If that will happen, Vorbeck will become a primary investment for me.

Graphene Nanochem PLC (LON:GRPH)

This company received this new name a few years ago. Before it was named Biofutures International PLC.

> *"On 10 December 2012, the Company announced that it had entered into a conditional agreement to acquire Platinum NanoChem Sdn. Bhd. ("Platinum NanoChem") which constitutes a reverse takeover under Rule 14 of the AIM Rules and is therefore conditional, inter alia, upon the approval of Shareholders at a General Meeting expected to be held on 25 March 2013.*
>
> *Following Admission, the merger of Biofutures and Platinum will result in an enlarged integrated business. The enlarged group is expected to be named Graphene NanoChem PLC.*
>
> *Platinum is a privately held global nanotechnology company whose business model is to design, formulate, manufacture and market a range of specialty chemicals and advanced materials (including products enhanced by a range of graphitic nanomaterials ("Graphene") which Platinum produces) that demonstrate improved performance characteristics when*

> *compared to existing products within the chemicals and materials markets.*
>
> *It has a revenue-generating range of high quality chemical products derived from readily available low-cost wastes from the palm oil industry." Source:* [(21)]

Since March 2013 the Malaysia based company Graphene Nano Chem has been listed on the London Stock Exchange. (GRPH).

What makes it already attractive is their existing business relationship with big players from the oil business.

They are capable of extracting Graphene from bio-degradable stuff (it seems mainly from the residues of the palm-oil production they own, too.)

Such, they can deliver large amounts of Graphene, which they seem to extract from bio-ethanol.

In my humble opinion, such businesses, among others, could make graphite mining companies obsolete.

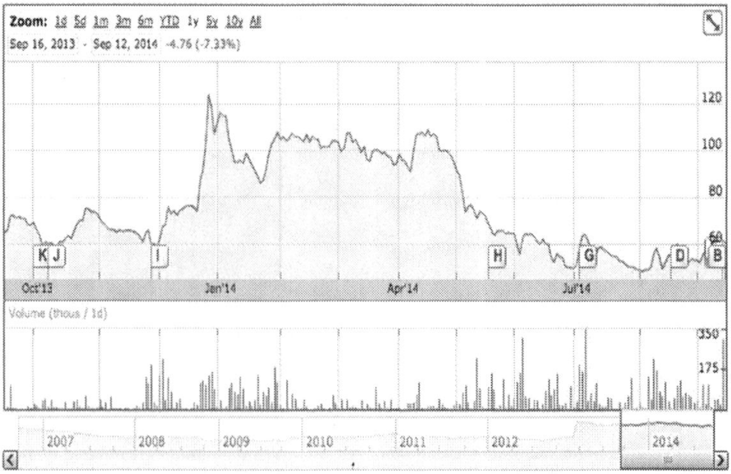

Graphene Nanochem stocks (source: Google Finance)

Since the merger their stocks have performed very poorly.

This does not prevent research analysts at Panmure Gordon in a report released on September 11, 2014 to give the company a strong „buy“ recommendation.

According to that analyst, the current target price of $2.43 of the stock has a potential upside of more than 140% of the company's current price.

XG Sciences

This company announced in April 2013 that it had invented a new generation of Graphene based material for batteries used in today's hybrid cars.

Their invention allows those batteries to store 4-times the amount of energy they could provide so far.

The current theoretical limit of battery storage capacity of Lithium-Ion batteries using Graphene is 10-times that of conventional Lithium-Ion technology in hybrid vehicles.

By quadrupling this power, XG Sciences are on the right way to come close to the mentioned theoretical limit very soon.

This company also offers additional Graphene core-products that are used by major companies.

Mr. Knox, their CEO, is known for selling startups successfully to larger companies.

Perhaps he and his management team will bring their baby to the stock market.

This privately held company involved in Graphene production with impressive partnerships (among them

South Korea market leaders Samsung and POSCO) and a great customer base should be observed carefully.

XG Sciences, Inc., formed in 2006 to manufacture and sell commercial quantities of a material that they call Graphene nanoplatelets. [62]

The College of Engineering at Michigan State University pioneered the manufacturing processes for this material.

Some of XGSciences intellectual property therefore is licensed from Michigan State University.

XG Sciences has 550+ customers in 32 different countries which include the likes of Boeing, Lockheed, Honda, 3M, and General Motors. [62]

They say, that their nanoplatelets could change the whole battery industry.

Therefore, it will be very interesting to follow their IPO, if it ever happens.

Samsung has a strong strategy focusing on the Graphene market and thus I'm suspecting that this Korean super-tanker will simply take over XGSciences, when the market will become stiffer than today.

CVD Equipment Corporation. [22]

Being a vendor of CVD systems and related material such as quartzware, they are probably a competitor of Aixtron.

They are listed on NASDAQ (CVV).

CVD stands for Chemical Vapor Deposition and is a process used mainly in the semiconductor industry.

They've been founded already in 1982, at a time, when the CVD market started to leave the science labs.

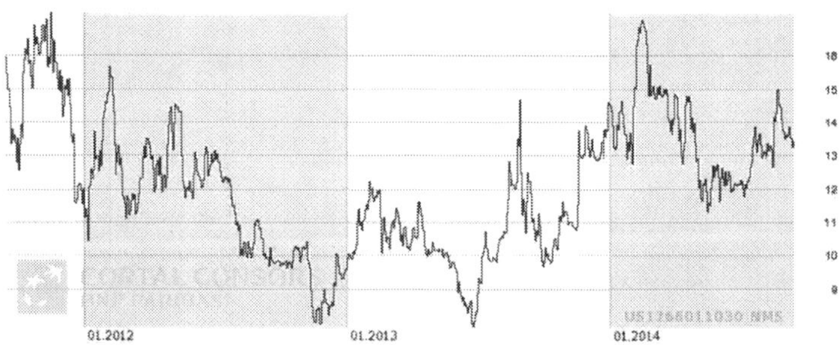

3-year stock performance of CVD Equipment Corporation

Given that the overall market situation for CVD companies such as CVD Equipment Corporation, Aixtron or VEECO has worsened over the last years, these stocks show a relatively good performance.

Over the last 9 years, CVD Equipment Corporation had been profitable and the shareholder return was close to +200%.

According to some analysts the reason for this is that CVD Equipment Corporation recently has invested into the Graphene market and such has lowered its dependence from the semiconductor industry.

If the commitment to the Graphene market had not been underlined recently by a strong message from their management, any stocks CVD Equipment Corporation would better be sold.

If the R&D efforts in Graphene become successful, CVD Equipment Corporations stocks are a clear "buy".

GRAnPH Nanotech (Grupo Antolin)

This company is interesting because it's part of the billion-dollar flagship project of the European Union.

With the funding of that project they should be able to boost their sales of Graphene core chemical products and invent additional solutions.

"GRAnPH® Nanotech is a trademark of Grupo Antolin, in the field of nanotechnology, with the objective of providing **superior quality Graphene** products for high tech applications, as well as other carbon based nanostructures and nano-composites." (Source: Grupo Antolin website)

They plan to become a leader in the market for high quality Graphene products and contribute to the development of new Graphene based electronics, energy devices and structural applications.

Customers are mainly from the electronics and semiconductor industry.

Knowing that Grupo Antolin is a major supplier for the automotive industry, I'm expecting them to invest in production of automotive components such as vehicle bodies.

BMW and Volkswagen have plans to build cars based on Graphene, and Volkswagen is already working with Grupo Antolin in some other areas.

Based in Madrid, Spain, Grupo Antolin continuously showed an excellent financial performance in the past, with revenues heading close to US$ 3 billion in 2014.

Crayo Nano

The Norwegian company develops and sells highly specialized hybrid products such as Gallium-Arsenide Nano-fibers on Graphene.

Their main market targets are: [23]

- Photovoltaics
- Light emitting diodes (LED)
- Thermoelectrics
- Piezoelectric
- 3D integrated circuits

Their products might find a ready market in the optoelectronics industry. Also not yet listed on stock markets, CrayoNano is another target to watch.

CrayoNano AS was founded in June 2012 in order to develop and commercialize the new hybrid semiconductor technology, and holds exclusive licenses for all relevant fields of use. [66]

It is on the radar screen for potential stock investments. As soon as an IPO is in sight, they should be a great opportunity.

Intellectual property is the highest ranked asset and currency in the fast moving, but still small, 2-D market. It seems that CrayoNano has abundant technology know-how.

This is making them a favorite investment target.

Angstron [24]

They claim to be the only company to be capable of producing high quality, consistent amounts of very pure or oxidized Graphene.

Obviously it is still a small company based in Dayton, Ohio. This company already holds an astonishing large number of Graphene related patents.

That's the interesting part: these patents might push Angstron's market value much above the value of its assets when it eventually enters the stock market.

Angstron Materials founders Dr. Bor Jang and Dr. Aruna Zhamu are ranked among the top five in the world for their development of intellectual property publications for Graphene. [23]

Gridlogics, a patent analytics solutions specialist that publishes Patent iNSIGHT Pro, ranked Jang with 42 and Zhamu with 39 Graphene-based IP publications. Such they were second only behind industrial giant Siemens.

Filings by Jang and Zhamu cover a range of new technologies, including new compositions, processes, and applications of both single-layer and multi-layer

Graphene, collectively referred to as Nano Graphene platelets.

Angstron Materials is the world's largest producer of Nano Graphene platelets. (Source: [(23)])

If you have a look at their website *angstrom.com* and search for the list of patents, you'll be impressed.

I assume that if they just started franchising their technological inventions, it would make Drs. Jang's and Zhamu's company a top player in the 2-D market.

If this company goes public eventually, it'll be one of the stocks to be bought.

AMO GmbH [25]

Based in Germany, this small startup also receives funds from the Graphene flagship project of the EU.

They offer a broad range of Graphene based products and solutions and also measurement devices as well as development for Nano- and optoelectronics.

Already in 2007 they could demonstrate a first Graphene based transistor.

That was at a time when many scientists doubted that such an electronic device could ever be manufactured.

AMO is located in the same town as Aixtron and might have learned from them the excellent performance one could achieve by entering the stock market.

On September 2, 2014 AMO published news about the development of a 50 Gigabit/Sec photodetector, based on Graphene.

50 Gigabit/sec is a world record and demonstrates the stunning capabilities of that material.

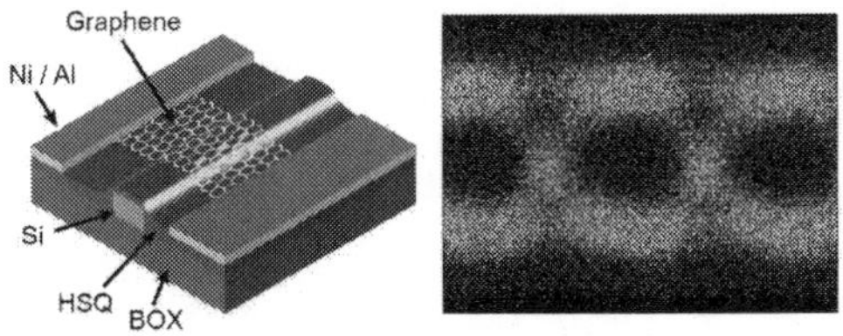

Picture: Graphic depiction of the 50 Gigabit/Sec photodetector (by permission of AMO GmbH)

AMO should be watched, not only because it seems to have the know-how to become a high flyer in the Graphene market, but also because they are part of the EU flagship project.

The latter will be a great financial "asset" for AMO, as they might be able to invest into more projects.

SouthWest NanoTechnologies Inc.[28]

Recently a venture capitalist named Dr. W.R. Bottoms was appointed to the board of directors for this Oklahoma based company.

For me that is a clear sign that they are preparing to enter the stock market.

Among their investors are NASA, US department of energy, Conoco Philips and some other institutions.

SouthWest NanoTechnologies Inc. (SWeNT) produces carbon nanotubes based on a patent they are holding.

It was founded in 2001 and is based in Norman, Oklahoma. In addition, it has a facility in Canton, Massachusetts.

SouthWest Nanotechnologies was created by Dr. Daniel E. Resasco, who graduated at Yale University and now works as a professor of Chemical Engineering at the University of Oklahoma. There he is head of the carbon nanotube research team.

Nantero Inc.

Based in Woburn, Mass., they develop semiconductor devices using carbon nanotubes.

Their devices include memory, logic, and other semiconductor products.

The company offers NRAM (Nanotube-based/Nonvolatile RAM), a high density nonvolatile random access memory chip; Nano-electromechanical switches; and microelectronic-grade carbon nanotube coating for bio-sensor and IR detector applications, and touchscreen display applications. [38]

They've received funding from venture capitalists, which makes a potential IPO already very probable.

Graphenea [26]

In the first edition of this book I viewed Graphenea with a lot of skepticism.

Since then I've followed their external communication and came to another conclusion.

They are not only a vendor of ultra-pure Graphene to many of the largest companies in the world.

This company also invests a lot of money into research.

Together with renowned research institutes such as the Max-Planck Institute in Germany, they've invented great applications for Graphene, such as multipliers for light power or water purification sheets.

The famous Sigma Aldrich corporation acts as their sales channel.

Recently they also published an invention where they doped ceramic, like Alumina-based sapphire, with Graphene.

The way they managed to develop this fascinating process showed that adding Graphene to that material increased its electrical conductivity dramatically.

Graphenea also started to investigate other ceramic material such as silicon-carbide, silicon-nitride, titanium-carbide, and Zirconia for similar properties.

Repsol, an oil and gas company with presence in more than thirty countries, employing over 23,000 people, recently acquired a major portion of Graphenea.

When Graphenea enters the stock market, they are a candidate for skyrocketing stock prices, for the same reason, CrayoNano or Angstrom are candidates.

This is their superior know-how in their specific niche market and focus on R&D.

Graphenea is located in San Sebastian, in the north of Spain, in the vicinity of the local university, which has some leading research facilities in the European Graphene science community.

XOLVE [35]

When I started writing this book, XOLVE [35] was on my radar screen for future newcomers, but I was skeptical that it could perform successfully on a large scale.

My perception changed when I learned of their new agreement with NTT Carbon Fiber Group of London.

Both companies agreed to engineer airplane component parts, such as carbon wings.

This agreement also acknowledges XOLVE's leading research capabilities.

Because XOLVE employs less than ten people, one might ask why they are eligible for investors.

Well, it seems that their patents and research capabilities may hold the key to a major breakthrough in 2-D material technology.

As I mentioned at the beginning of this book, 2-D material is not a simple bulk material.

In fact, it is very difficult to handle and one has to invest a lot of effort in order to keep the 2-D material, as it should be.

In this respect, 2-D material is as complex in manufacturing as any nanomaterial.

Dr. Allen Clauss, Vice President of R&D for Xolve, Inc. recently published a statement about the potential pitfalls of 2-D nanomaterial, where he gave an overview about the potential traps awaiting any newcomer. [48]

To make a long story short, Xolve holds the patents, which could dramatically reduce development as well as manufacturing costs for any Graphene and 2-D material production.

Their solvents are also capable of extracting Graphene from simple graphite, further reducing the production costs of 2-D material.

With the chemical solvents invented by XOLVE, Graphene and other 2-D material could be produced on the competitive scale necessary to provide for large manufacturing companies.

Launching Graphene business beyond small-scale electronics could then happen at a rate much faster than scientists around the world originally suggested.

The current size of Xolve makes it an excellent choice, and president and CEO John Biondi seems willing to accept serious investors.

For small-scale investors, such as the common high-value stock picker, patience is required until this company goes public.

But then XOLVE has the potential to outperform many other stocks in the same area.

Warning about faked investment opportunities

At this stage of the book I want to raise awareness for fraud related to investments in the Graphene market.

In 2013 the FCA, the financial watchdog of the UK's stock market, issued a warning over investing in 'miracle material' Graphene.

Company's shareholders and private investors had received unsolicited phone calls concerning investment matters.

The FCA warned consumers about being sucked into "dubious" investments in Graphene amid excitement generated by the so-called "miracle material". [57]

Graphene investments are not "dubious" by origin. On the opposite, they are highly attractive, even though somewhat risky.

What are shady are the offerings of unknown companies claiming to know about investment opportunities in that market.

The FCA said:

“We believe the same firms that sold other high-risk, dubious products such as carbon credits, rare earth metals and overseas land and crops now try to sell Graphene.

“We are yet to see any convincing evidence that there is a viable market for retail investors to make money from investments in rare earth metals.

“There is a strong possibility of fraud with Graphene because it is unregulated and it is difficult to confirm that you have bought the genuine product.” (58)

The FCA continues saying:

„Shareholders are advised to be very wary of any unsolicited advice, offers to buy shares at a discount or offers of free reports about the Group.

The Financial Conduct Authority ("FCA") provides the following guidance should you receive unsolicited investment advice:

- Make sure you get the correct name of the person and organization who has called; and
- Check that they are properly authorized by the FCA before getting involved with them. You can check to see if they are registered with the FCA at www.fca.org.uk/.

The FCA also maintains on its website a list of unauthorized overseas firms who are targeting, or have targeted, UK investors and any approach from such organizations should be reported to the FSA so that this list can be kept up to date and any other appropriate action can be considered.

If you deal with an unauthorized firm, you would not be eligible to receive payment under the Financial Services Compensation Scheme. [(58)]

As with all investment opportunities, freeloaders and dubious companies are a sign of a market offering opportunity.

Anyhow, one should not let suckers fool one with false promises and sometimes bold lies.

As somebody who wants to participate in the huge prospects, which are to be expected in the 2-D market, you got to check carefully the companies you want to invest in, on your own.

Eventually there will be credible stockbrokers offering funds for the Graphene and related markets. But those will be the same ones offering serious stock investments today.

A few brokers investing in "Green" stocks also are going "long" today for Graphene investments.

Check out respectable sites like Forbes.com for those, if you decide to work with a broker.

For the time being, one has to control every detail (except for companies like IBM or other major players).

This might become tedious at the beginning, but – as with Microsoft in the 1980s – it'll be a chance to own the right stocks before anybody else has realized what's going on.

Other companies to observe (with a grain of salt):

Nanoridge [27]

2-D Tech [29]

American Graphite Tech. (NASDAQ: AGIN) [30]

Bluestone Global Tech [31]

CAL Battery [32]

Durham Graphene Science [33]

Garmor Inc [49] *new to this list*

SiNode Systems [36]

Vulvox Nanobiotechnology [34]

Planartech [63]

Graphene Industries [65]

These companies might deserve better recognition eventually. But as of the time I'm writing this book, I don't see an added value for investors, so writing about these companies is not required. When some interesting information will emerge one day about any of these operations, I'll start investigating further.

XOLVE had been on this list before. But since they've been sending out compelling information to the public, they deserved a closer look and I added them in the section about Graphene investment opportunities.

Chinese companies

There are plenty of companies in China that are involved in some kind of Graphene business. For instance Nanjing JCNano produces a wide range of Graphene related material.

CNano and Timesnano are vendors of Graphene as base products, so one should expect them to become players in the application area as well.

China itself ranks second in the world in Graphene related patents. (The USA is leading this list).

But up to this time I haven't found any startups that could become publicly traded.

Anyhow, China is an area one needs to watch carefully and patiently.

According to some analysts, China's Graphene industry is not only well underway, but will skyrocket in 2014.

China has become the world's biggest developer of the Graphene sector.

The "Sixth Element" is a Chinese company located in the Changzhou area. In 2013 they announced that they started mass production of high-quality Graphene powder with an annual 100-ton capacity.

On December 20th, 2013 China's Ningo Morhs Technology announced that it had begun operating the world's largest mass production line of high-quality Graphene material in Ningbo City.

The plant announced providing an annual capacity of 300-ton of Graphene per year in the first phase, and after finishing its second-phase project, it plans to have an annual production capacity of 1,000 tons of Graphene material.

This marks the beginning of a new era for low-cost mass production of high quality Graphene, as this material is becoming an industrial raw material of large supply. (67)

There are more announcements of that kind and the Chinese government does everything, which is possible to encourage mass production of Graphene by means of chemical processing, not mining.

This strongly underlines my argument that mining companies are on the verge of becoming obsolete.

Large companies to observe: [14]

BASF, Bayer, IBM, Samsung, Sandisk,

Basically, these are the companies which have announced working in the field of Graphene and those which are receiving heavy funding from the EU Graphene flagship project.

Let's have a look at each company:

BASF:

This is one of the leading chemical companies in terms of research of Graphene. They've aligned forces with the famous German science institute "Max-Planck-Society" in 2012 and are pushing carbon (i.e. Graphene) applications.

Most of the Graphene related research was not done at BASF labs itself, but at their partner company Vorbeck Corp., based in Jessup, Maryland.

Vorbeck is working with Princeton University on new products based on Graphene.

I wouldn't be surprised if BASF pushed Vorbeck onto the stock market eventually, but even having them as a subsidiary might be enough to let BASF's shares rise.

Bayer:

This chemical company is capable of producing large amounts of Graphene as a core product.

As this product is still difficult to produce in high quality, BASF should be ahead of the rest of the industry within 2-3 years.

Given the nature of that company, Graphene will play only a small role in their business.

Such it is improbable that their stocks will follow the exponential growth curve of the Graphene stock market.

IBM:

Along with Samsung, this is the most interesting company in terms of technology.

During the past decades that company has led any technological development in the area of microprocessor and semiconductor technology.

Their management is known for making visionary decisions and they've not only survived the turmoil of the computer industry, but came out of any crisis stronger than before.

C.Y. Sung, one of their R&D managers recently said in a presentation, that by 2020 they'll have completely new non-CMOS devices, technologies and manufacturing processes "with increased performance and density for several orders of magnitude beyond the limits of CMOS". (17)

(CMOS is the acronym for the process technology used in today's semiconductors)

It is not surprising that IBM last year announced a record braking Graphene transistor. [15]

Its speed is a stunning 100 Gigahertz (cutoff frequency). That is much more than any silicon transistor is capable of providing.

IBM fellow Phaedon Avouris stated that they are looking to target the Terahertz range with new devices.

That's another increase of the transistor speed by a factor of 10 within a short timeframe. [16] And this is only the start of the race.

At the same conference where C.Y. Sung spoke about the Graphene transistor, they also demonstrated a Graphene based photo device in an optical communication link, Graphene thin-film sensors, super capacitors for high performance battery cells and solar energy devices. [17]

IBM with its financial power and its world leading R&D labs has been able to show the first products in semiconductor and other technology areas related to it.

This is a proof that the Graphene era has come already. In the long range IBM might be the first company to be able to produce a quantum computer in their "Watson project".

IBM is on the list of stocks I'll buy.

Samsung:

Samsung's "Advanced Institute of Technology", their R&D arm, recently announced some breakthrough in developing advanced transistor technology based on Graphene.

It seems that this Korean company is as much at the forefront of scientific development as IBM.

As Samsung is selling more into the consumer market, they'll probably use their knowledge to produce smaller processors and also optical devices replacing or improving OLED and other stuff currently used for TV etc.

SanDisk:

As for Samsung and IBM, Graphene research today is only one of many fields Sandisk is investing in.

They hold about 30 patents for Graphene applications. Therefore, they are one of the major players in this field.

The Milpitas, California, based company sees the future of its flash memory products in Graphene instead of silicon.

By leading the field in R&D for that specific product group, they probably will outperform any other memory storage vendor.

Flashcards for computers or cameras could store many Terabytes, enabling computers and optoelectronic devices one never thought about.

Others:

Texas Instruments, Micron and other semiconductor companies will benefit from public research initiatives such as the NRI: Nanoelectronics Research Initiative. (37)

Within the "Semiconductor Technology Advanced Research Network" (STARnet) (44) they are deeply involved in further research.

Texas Instruments sees Graphene as a potential candidate to further extend Moore's Law for decades to come. The material shows great promise for future thermal management and energy storage devices, those that require operation at very high frequencies, and many more. (43)

Their main research areas are about using Graphene for flexible electronics, sensors, thermal spreaders, touch screens, transparent conductive electrodes, photonics, high frequency electronics and others.

How will Graphene technology evolve over time?

2014: The base technology is established, first gadgets to see

The first devices of transparent and conductive Graphene films are available at reasonable cost. Flexible electronics and optoelectronics devices are on the brink of entering the market.

Work on printed Graphene RF-tags (such as the price tag on a book cover), Graphene supported foldable OLED (*organic light-emitting diode*) screens, and flexible e-paper is leaving R&D laboratories and test devices are available.

First touch screen and tunable displays might be able to be demonstrated at fairs.

A Chinese company. AWIT Inc., has already sold its first batch of 2.000 Graphene based smartphones.

Small companies to observe:

+ISO

AMS Technologies

AsTEK

C3Nano

Inteltronic

ChemPacific

CreaPhys

Dresden Microdisplays

AU Optronics

Advantech

Cambridge Display

CSEM Switzerland

COMEDD Germany

Yunnan North OLiGHTEK

Big players:

Dupont, BASF, Bayer, Sony, Samsung (with its affiliate Cheil Industries), IBM, Henkel, Kimberly-Clark, etc.

I'm keeping stocks of Bayer, BASF and IBM mainly for the reason that I see them as a reliable payer of dividends and being a key player in the IT or chemical industry.

IBM and Samsung might be the stocks that have the capability of exceeding the rest.

Other companies such as Siemens, Mitsubishi or Exxon, just to name a few, might also play a role in developing core technologies.

As for Siemens, one might argue that this company holds a huge number of patents, which would make them a market leader in the Graphene world.

Siemens in the past had consistently and repeatedly overslept many new exciting opportunities.

Be it mobile phones, the laser printer (initially a Siemens patent), automation software and more.

Seeing that their management hasn't learned very much from the past, I always would go "short" on their stocks.

2016: Optical technology, lasers, communication hardware

Graphene enters the communication market. Mode-locked lasers, modulators and photo detectors are going to speed up fiber-optical communication systems. The first prototypes of foldable wrist phones and ultra-light tablet PC might be seen.

<u>Small companies to observe:</u>

Trumpf

AMS Technologies

Boston Applied Technologies

VEECO

...

(*Just to name a few*)

There are 200+ companies in this market area. A way to get a grip on the necessary information is by observing the LFW Optoelectronics Market Financials (LFWI) index.

Not all companies listed in that index are capable of jumping onto the fast accelerating Graphene and 2-D train, since only a few seem to provide the financial power needed to invest in the technology changes.

Big players:

Apple, Google, Nokia, Texas Instruments, EADS, Boeing, BMW

Yes, I see **Apple** and **Google** as key-players in that market by 2016. They have the capability of building new products.

Both companies are also attractive for alumni from engineering schools, are fascinating to scientists, and also have the financial power to stand a risky and unforeseeable development. And they have managers with visionary genius.

I would not be surprised if the first company to throw a foldable mobile phone into the market would be Apple.

Google with its glasses project is also on the right way. If they just swallow a small startup that works on Graphene, for them it will become a piece of cake.

Nokia in 2008 showed a video featuring a mobile foldable wrist phone.

By that time they had no idea how they were going to make that gadget happen.

Their partner research institute in Cambridge, UK, made them aware of the stunning features of Graphene and then it became obvious how their vision could lift off.

So far it looked as if Nokia would be the first company to market this fancy innovation.

Nevertheless, Nokia is not a company I shall invest in.

The reason for my negative attitude is simple: at the time being, they don't have more than a vision to show, and even that vision has not been developed further during the last years.

In addition, they've demonstrated in the past that in difficult times they are willing to toss intelligent ideas overboard for the sake of cost reduction.

In the year 2007 I read in an article of "Financial Times" that they'd outsourced their R&D to low cost countries, for me it became obvious that this firm would face a hard time and one should stay away from their stocks. (45)

R&D is the heart of any technology company and giving away the knowledge accumulated within the heads of the R&D staff comes close to a death warrant.

Many mediocre and sometimes erratic decisions of the management of this enterprise indicate to me that they seem to lack the financial power as well as the spirit which is necessary to accomplish the complex and risky technical investment decisions required to market their vision. [6]

Maybe, their management should return to their historical products. Those were gum-boots and those will be manufactured with Graphene one day as well.

There are two or three enterprises in the market that understand to launch fascinating new products in the consumer area.

These are e.g. Apple, Samsung and may be Motorola, now a Google business. These are the ones I would bet on.

(When I wrote this chapter for the first time, Nokia Mobile Business was still owned by Nokia itself. Shortly after Microsoft acquired that business division.)

Both **Boeing** and **Airbus** will need to invest in Graphene in order to maintain the lead in building planes.

It's for a reason that Airbus entered the Graphene flagship project of the European Union with its 1.3 billion dollar funds.

Graphene offers capabilities needed urgently by all makers of planes.

It is ultra-light, stronger than any other material, thermally absolutely stable and conductive.

China has invested a lot in its own aerospace industry and they are also investing in Graphene research.

This is the biggest threat for Boeing and Airbus: Chinese companies will bring their Graphene plane to market before them.

Airbus Concept Plane (courtesy: Airbus SAS)

Airbus is a good investment, when it will be taken out from the EDAS group and stocks are offered on the market.

As EADS is chronically sick because of lack of cash, I'm expecting that they'll eventually take that step.

The same is valid for **Boeing**: if they separate military and civil business, their stocks might rise again.

BMW has gone through a fascinating transformation during the last 10 years.

They not only streamlined their production line and returned to building fancy cars. In addition, their engineering process went through a transformation.

As a result they are now able to flood the market with electric cars.

They are capable of producing hybrid cars (electrical/fuel powered) as well as using advanced fuel cells.

But this is true for many other automotive OEM such as GE, Toyota, Honda, Mercedes and so on.

BMW some years ago with their joint venture partner SGL Carbon started a carbon-based facility in Moses Lake (Washington) where they produce non-crimp carbon fibers.

This will enable extremely light and at the same time extremely strong chassis.

The know-how they gained from that investment might make them the only company being capable of switching to Graphene based chassis before 2020. Carbon fibers and Graphene are based on the same technology and this is unique to BMW.

BMW i8 with carbon technology (courtesy BMW AG)

Graphene is not only much lighter than steel, it is also much stronger. Even roofs for cooling or heating a car could be built by using the same material.

I see BMW as the first company capable of engineering and bring to market a sports car needing less than 0,5 Gallons/100 miles.

2020: Bio-Pharma and Automation Startup Investment opportunities are visible

This is the year when investing opportunities will become really interesting. Sensor networks for food quality detection, biological sensors, DNA sensors and first medical devices built with Graphene will enter the market.

Even the first prototypes of hybrid biological/electronics computers using DNA for computing and direct interfaces to nerve cells are thinkable.

Scientists from MIT and Harvard in April of 2013 published a stunning success story about building DNA based electronic devices on Graphene. [41] So this vision is based on reality.

The nerves in the spine of paraplegic people might be bridged by similar devices.

By then the **bio-tech industry** might be back into the stock market in the way it performed at the end of the 1990's.

Companies to observe (bio/pharma):

Qiagen

Amgen

Gilead Sciences

ARIAD Pharma

Seattle Genetics

Biogen (I am an owner of their stocks)

Merck KGaA

Baxter

Actelion

Regeneron

Forest laboratories

Celgene

Elan

None of these companies, except Quiagen, are small. So it will be worth watching new companies entering the stock market. Biotechnology is an area where a clever scientist could easily get out of the laboratory and found a new company with the help of a hedge fund within a short period.

Qiagen recently published a paper where they revealed research on Graphene nanosheets, which could make them a leader in using that material for medical applications. [39]

Prof. Henk Postma, a physicist from California State University, Northridge, thinks that Graphene could be used for sequencing DNA.

Robotics technology will be another area dominated by Graphene applications in 6-8 years from now, and we'll see breathtaking gadgets.

Micro- and nano robots could be built completely from Graphene.

As that material also is a power provider, these systems could work completely independent from any power source.

Also other self-powered, flexible mobile and stationary devices are thinkable.

Companies to observe (Robotics):

ABB Robotics

Adept Technology

AeroVironment

Aethon

Aldebaran Robotics

Astrobotic Technology

Barrett Technology

Bluefin Robotics

Bossa Nova Robotics

Boston Dynamics

Clearpath Robotics

Cyberdnye

DENSO Robotics

EKSO Bionics

Energid Technologies

EPSON Robots

FANUC Robotics

FESTO

FMC Schilling Robotics

Hansen Medical

General Electric (medical robots)

Honeybee Robotics

Hyundai Heavy Industries

Innovation Associates

InTouch Health

Intuitive Surgical

iRobot Corp.

John Deere Mobile Robotics

Kiva Systems

KUKA Robotics

Lely

Liquid Robotics

MAKO Surgical

MathWorks

Medrobotics

Northrop Grumman

RedZone Robotics

Rethink Robotics

Robosoft

Robotiq

Schunk

Seegrid

Stäubli

Swisslog

Universal Robots

Vecna Technologies

VGo Communications

Willow Garage

Yaskawa Motoman Robotics

Many of the companies mentioned above are privately owned. Nevertheless, I found it worth displaying them all.

When any of these firms will go public, one should have a close look at it.

For instance, the company Mathworks: it's a privately owned software company located near Boston, Mass.

Unknown to most people, its core product Matlab is used in more than 90% of all cars, planes and robotics as well as medical applications.

This company is a market leader in control software required to run any complex machine. When this company eventually will go public, their stocks should be a prime selection.

2024-2030: Graphene is established

By now Graphene and other 2-D technologies have taken the lead in conquering the market. Light-weight batteries, super-capacitors, fuel cells and any kind of solar power devices will bring dramatic changes.

Solar cells will not be heavier than a piece of paper and will be attached to roofs with the help of Graphene based super-glue. Graphene wires bring the power loss-free to an energy storage battery.

It is no longer necessary to extract hydrogen in an expensive physical process.

Graphene and Boron-Nitride membranes should be able to get the basic element for our future cars out of water with the help of a small amount of electricity.

Prostheses and artificial retinas have become state of the art in bio-medicine and glasses are nearly invisible as contact lenses are integrated in the eye.

Books and TVs return to their initial state – they are thin as paper. Computers will resemble todays tablet PCs only to the extent that they have a similar form factor.

Food can be grown in farming towers in towns directly, as Graphene membranes can extract water from even the smallest amount of humidity in the air.

At the same time Graphene will deliver the power to heat or cool simply from solar energy.

Airbus and Boeing along with other plane manufacturers should be able to fly first prototypes of large but ultra-light planes made of Graphene.

Cars are also made of that material and will only use batteries and fuel-cells made of Graphene as well.

But the most interesting development might come from an area unforeseen to most scientists and business gurus today.

As said above, Graphene does not allow any gas – not even helium – permeate through its surface. Only electrons can pass through, if directed onto the ultra-thin film.

Any physicist immediately will understand the impact of this effect.

Electrons behave, like all particles and also some molecules, in 2 different ways, when moving through a grid such as a Graphene film. They can either be detected as a wave or as a particle.

Since Graphene can control the flow of quantum particles, we might use it to build quantum computers, which do not only use 2 digits – 1 and 0 – for computing, as today's systems do.

There could be microprocessors using a large number of computing states. If, let's say, we go from a binary system to using the numbers 0-9, i.e. 10 digits, the speed of computing will increase not by a factor of 10 but by many billions of operations per second.

Some scientists already have shown concepts demonstrating a way how this could be achieved. (18)

Also complete new experiments in the scientific trenches, e.g. in large colliders such as CERN in Switzerland, are coming into my mind.

High-speed quantum physics could make expensive power consumption experiments obsolete.

Experiment at CERN in Geneva/Switzerland: Graphene based supercomputers will change how physicists explain the world [P-13]

Here it is where the circle begins to close – but on a higher and more complex level.

60 years ago the first silicon based semiconductor devices entered a perplexed market.

Nobody seemed to know what to do with these products at that time. They were considered as expensive, useless and exotic.

Today we know that semiconductors changed the world and our lives in a way not seen before in the history.

The main reason for that success has to do with quantum physics. Semiconductor effects are based on quantum effect.

By understanding this mechanism and exploiting its capabilities on a macroscopic scale, scientists and engineers had been able to develop the most sophisticated technologies possible.

Everything we knew at that time changed in the most dramatic way.

Graphene has the potential to change our world again in such a manner.As of today, I can't see any new company on the horizon that might take the lead in developing Graphene based technologies. But this will become the most fascinating area to observe.

New developments

Many companies and research organizations are investing a large amount of money and manpower in this challenging new field of 2-D material science.

A furious race between competing teams is already paving the way to patented products.

This chapter highlights new developments in the area of 2-D material as well as interesting research results and companies of note.

Digging through a lot of publications and websites I often found reports, which doubted that the Graphene market ever would become more than a hype.

It was hilarious for me to read a serious blog post or an analyst statement from one of the deniers of the Graphene success story on one day and find the allegations and conclusions of these people being stultified on the next day.

One day I read an article by a credible university professor who claimed that there never would be any microchip or even a transistor based on Graphene.

He reasoned the missing “band-gap” (a physical property needed for semiconducting behavior) in Graphene would never make that material useful for electronic devices.

The week after, during which I’d wondered whether this professor might be correct, IBM published a fascinating article, presenting not only a transistor but a complete microchip made of Graphene.

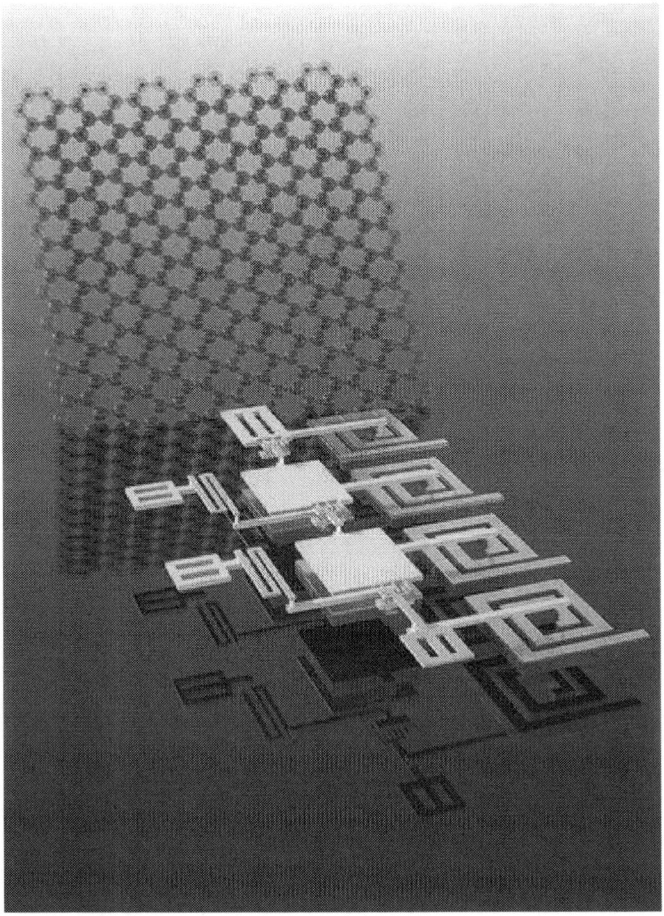

IBM microchip based on Graphene (source: IBM, by permission [68]

The picture above is not showing the initial development but a succeeding microchip, presented to the press by January 2014.

> *"Researchers have built the world's most advanced fully functional integrated circuit made of wafer-scale Graphene – a novel semiconductor material that has the potential to improve today's wireless devices for cheaper, high-speed communications. The nanotechnology milestone opens up new carbon-based electronics device and circuit applications beyond what is possible with today's silicon chips."* [68]

What a Eureka moment!

Similar occurrences happened over and over again. The scientific development is moving at such a fast pace, that even professional members of the community aren't capable of staying at the front of the wave.

So one should expect that the more or less anecdotal reports below were outdated already by the time of reading.

There was a press article, where solar cells based on Graphene were reported. It was proudly announced to have some 5% of efficiency.

The week after I'd written about that in this book, another researcher published a scientific report showing that Graphene based solar cells could already achieve an efficiency of 15%. Sic!

Some years ago half of the professional solar cell scientist were convinced that Graphene based solar cells never ever would be possible (see the missing band-gap phrase).

Below you'll find several stories, which are meant to highlight recent developments.

Digging through similar articles will help an investor to develop some kind of a Graphene investment "gene" or intuition, as it will make anybody familiar with the application.

If you read through this, there is no need to understand more than basic terms of technology.

Some of the phrases used in these articles could eventually come handy.

May be you could impress a person with your knowledge at a party? Graphene related information would leave many people with gaping mouths. (I'm just kidding...)

Tesla's Elon Musk becoming "superman" through "supercapacitors"

Being able to travel less than 270 miles on a single charge, the range of the batteries as well as the time needed to recharge the batteries of a Tesla car discourage many potential buyers of Tesla's vehicles. In order to compete with most of the cars on the market today, electric vehicles must be capable of traveling 500 miles and recharge within a few minutes only.

Graphene can be used as the basic component of a "supercapacitor". This is an electrical power storage device combining the benefits of a battery (long time storage of electrical power) with that of a capacitor (fast recharging cycles).

A supercapacitor charges far more rapidly than chemical batteries. Graphene's structure also offers a much higher energy density than supercapacitors based on other material. It can hold a vast amount of electrons, therefore outperforming any battery in the amount of charge it can provide.

A team of engineers at Northwestern University in Evanston, Illinois, has created an electrode for lithium-ion batteries (rechargeable batteries), allowing batteries to hold a charge up to 10 times greater than current technology. Batteries with the new electrode also can charge 10 times faster than current batteries.

The researchers combined two chemical engineering approaches to address two major battery limitations -- energy capacity and charge rate -- in one fell swoop.

In addition to offering better batteries for cell phones and iPods, the technology could pave the way for more efficient, smaller batteries for electric cars.

Another factor 10 is allocated by the extended lifetime. Even after 150 charges the new lithium-lon battery with Graphene anodes is far more effective than lithium-ion batteries used today.

Just to remember that this had been invented already in 2011.

By June 2014, Prof. Richard B. Kaner from University of California in Los Angeles added new investigations to those results, showing that, as a next step, future supercapacitors could be using Graphene only, replacing Lithium and other material.

According to Elon Musk, the CEO of TESLA, the technology invented by Northwestern University will be used in future versions of Tesla's cars.

This company could soon offer the much-wanted 500-mile battery, based on Graphene-based anodes, which can quadruple the density and output of lithium-ion batteries.

But this should be an intermediate step only.

One can assume that within 3-4 years Prof. Kaner's technology will be mature enough to improve the technology of supercapacitors further.

At that time Tesla's cars might reach the competitive breakthrough, which will let them outperform traditional vehicle manufacturers.

Graphene is exorbitantly expensive at the moment, when compared to e.g. Lithium. But one has seen dramatic plunges in the price of that suspected wonder material already.

Given the dramatic progress in the production technology for Graphene and other 2D material, by the time of the maturity of the battery technology Graphene will be available in excess on the market. Prices for that base material will drop exponentially.

Therefore, I’m not surprised that Forbes listed Tesla as one of the upcoming hotspots of investment.

Stock manager Morris, who is co-founder of the Ecological advisors fund, keeps shares of Tesla in his Epiphany Global Ecological Mutual Fund EPENX. [69]

AMO: Graphene-based photo detectors outperform existing technology

German based AMO GmbH (Aachen,) reported on September 3, 2014 that they had created the world's fastest photo detector, operating at a maximum data rate of 50 Gbit/s.

Using Graphene as the most promising material for future ultrafast and super compact telecommunications and optical systems, they achieved this progress within a stunningly short timeframe.

Using that method, they were able to demonstrate the simplicity of integrating the new technology into mature state-of-the art silicon based systems.

Together with researchers from French telecoms giant Alcatel they used a conventional silicon photonic platform for designing ultrafast data communication technology systems.

Using conventional chemical vapor deposition (CVD) the consortium grew the Graphene on conventional silicon photonic waveguides.

Graphene music device or just a quantum dot?

This is not a hoax. Scientists created small nano-scale drums in Graphene, which turned out to be a great application for e.g. sensor in mobile phones.

Researchers from TU Delft's Institute of Nanoscience published on August 24, 2014 that they can observe extremely small changes in position and force on very small drums of Graphene.

Those drums could also become memory chips in a future quantum computer system. This became apparent when the focus of the scientists turned to the mechanical properties of the 2D material.

Because of their extremely low mass, one could experiment with tiny sheets of Graphene, as it would be the drumhead of a musician. Instead of drumsticks, the team used microwave-frequency light to 'play' the Graphene drums, to listen to its 'Nano sound', and to explore the way Graphene moves in these tiny holes.

Using a 2D crystal membrane as a mirror in an 'optomechanical cavity', Dr. Singh and his team were astonished to find the new properties.

In this experiment, they directed microwaves at a nano-scale Graphene drum. The drum acted as a mirror, reflecting the incoming waves and by that generating an interference pattern.

This interference pattern of the microwave photons bouncing off of the drum revealed changes in the position of the Graphene sheet on a scale of only 17 femtometers. This is about 1/10000th of the diameter of an atom.

One of the most intriguing results of that scientific project was the finding that one could use these drums as electronic memory devices for microwave photons.

By that photons can be converted into mechanical vibrations and stored for up to 10 milliseconds. That is not long by human standards, but it is a long time for a computer chip.

To study quantum motion is one of the next goals the research team has focused on. They want to investigate how oscillations on a quantum scale do appear. If one hits a classical drum with a stick, the drumhead vibrates in an observable pattern.

If one would film the drumhead with a high-speed camera, the membrane would be seen swinging between an up and down position.

That would be different in a quantum drum. The drumhead would show a 'quantum superposition'.

(Describing the principle of a 'quantum superposition' is not within the scope of this book, which is targeting investors, not scientists. There are great articles available on the Internet (e.g. Wikipedia) about this stuff. I recommend reading those, if you want to understand the related physics.)

This quantum motion (initially described by the physicist Erwin Schrödinger in 1935 in his famous thought experiment, featuring his famous "cat", is not only of scientific relevance.

Used in a quantum computer, this technology would act as a quantum memory chip.

This would be another step onto affordable quantum computers. (70)

Carbon nanotubes introduced into the next generation of solar cells

Carbon nanotubes (CNTs) have long been considered for photovoltaic, replacing conventional silicon.

This material is much lighter, cheaper, more flexible and less prone to long-term failure than traditional materials.

For a while taken into account as the basis for a future solar power device, research efforts halted some years ago when it became obvious that simple CNTs are by far not efficient enough in terms of converting sunlight into electrical power.

One might ask what carbon nanotubes (CNTs) might have to do with Graphene. Well, these CNTs are nothing but small sheets of Graphene curled up into tiny tubes.

A research team led by Mark Hersam, professor of materials science and engineering at Northwestern University's McCormick School of Engineering, announced on September 3, 2014, a new type of CNT solar cell that is twice as efficient as its predecessors.

It is also the first CNT solar cell to have its performance certified by the National Renewable Energy Laboratory, as Prof. Hersam stated proudly when presenting the product.

With an efficiency of 3% only these CNTs still lag behind conventional silicon solar cells, which show efficiency rates above 15%. But silicon devices are much more expensive to manufacture.

Prof. Hersam believes that he can achieve the same efficiency as that of silicon by stacking multiple layers of CNTs where each is optimized for a particular portion of the solar spectrum and capable of absorbing more light by this.

Incorporating other materials, such as organic or inorganic semiconductors, to complement CNTs, might be another path to follow.

This might feed the arsenal of arguments of people denying the potential of Graphene and CNTs for making solar power devices.

But those people should try to remember the early stages of development of silicon solar cells. At that time, scientists considered efficiency rates of 3% to be a great success at the beginning.

So I'm convinced that clever engineers and scientists soon will find ways to copy the dramatic leaps we've seen in silicon systems in the past and make Graphene based solar cells a reality.

When I finished writing this, I got news from a research team that was about to publish a scientific article, in which they would announce reaching an efficiency of 15% for a Graphene powered solar cell.

The gap between silicon and Graphene performance seems to close in an astounding fast speed.

I didn't change the related article, as this demonstrates what's happening in the Graphene research field currently: new developments are reported in an exponentially growing amount and it is impossible to keep track of every new finding.

Defect-free Graphene required for Li-ion batteries

There might be better news for Tesla's Elon Musk, supporting his project to introduce lightweight, high performance batteries into Tesla's cars.

A research team from Korea Advanced Institute of Science and Technology (KAIST) and a research group from Hanbat National University in Daejeon, South Korea, announced in August 2014, that they've detected a process that allows them to create defect-free Graphene.

This might become a big game changer for the battery industry, as it will become possible now to wrap sheets of negatively charged defect free Graphene around positively charged Co_3O_4 in order to create an anode for high-performance Li-ion batteries.

So far all Graphene produced by conventional methods showed defects in the sheets and it had been very difficult to create even small amounts of defect-free Graphene.

Industrial applications - on the other hand - require chemically pure, structurally uniform, and customizable base products.

This kind of Graphene now becomes affordable by using the methods described by the Korean research team.

Currently there are two methods used to fabricate high-quality Graphene: mechanical and chemical approaches. While it is possible to produce high-quality Graphene using mechanical methods, the low yield of that process renders it insufficient for large-scale production.

By means of chemical methods, one can produce large quantities of Graphene. But those can often show imperfections.

The researchers used another approach, very different from what is standard today. By melting graphite (which is the 3D form of 2D material Graphene) in conjunction with Potassium, they were able to extract defect free Graphene.

One of the experiments showed that the Graphene sheets produced by this process created outstanding electrical features when used as an anode in Li-ion batteries. Its capacity outperformed all known electrode material.

This is highly relevant for investors, since lithium-ion batteries - until the advent of pure Graphene supercapacitors - are critical not only for electrical vehicles. Companies like Apple or Samsung are starving for new small Li-ion batteries for mobile phone.

Their wish for high power batteries in customizable sizes and with lower prices could be fulfilled soon, when the industry will have introduced the new defect free Graphene into their process.

As mentioned before, investors need to watch out for companies that are capable of switching their production facilities for using Graphene instead of Lithium.

In four to five years Graphene supercapacitors should be able to replace Li-ion devices.

Graphene oxide fibers

Known under the trade name Kevlar, this high strength material is a well-known synthetic fiber made of para-aramids used for replacing e.g. steel or building lightweight body armor.

Prof. Mauricio Terrones from Penn State university recently announced that his team had found a potential replacement for the aramid fibers mainly produced by DuPont industries.

This new product is based on Graphene-oxide, a derivative of Graphene.

Despite being a form of carbon, Graphene itself does not oxidize. It is chemically inert and oxygen needs to be forced into the hexagonal grid, which Graphene is made of, by some kind of a chemical or mechanical process.

Prof. Terrones' team invented a simple method of producing strong stretchable Graphene oxide fibers. Even the first samples of the new product show strength similar to those of Kevlar.

By adapting the amount of oxygen atoms within the Graphene layers it is possible to tune the electrical and optical properties of the yarns of Graphene-oxide.

This adds new features to the mechanical strengths of the fibers, opening a wide range of potential applications.

The Graphene-oxide fiber could easily be molded into car tires, for instance. Their ability to transport electricity, then could help making them sensors at the same time.

Fast-speed - low-cost photonics with Graphene

Professor Baohua Jia and a team of researchers from Swinburne's Centre for Micro-Photonics recently announced development of a high-quality, continuous Graphene-oxide thin film that could eventually enable ultra fast telecommunications.

They created an ultra-thin film with astounding optical properties, suitable for high performance integrated photonic devices used in optical communications, biomedicine, photonics computing and more.

It may be considered for use in future quantum computer systems as well, as those super-fast systems rely heavily on fast inter-processor communication systems.

According to the team member's opinion, they have found a new platform in which they can produce each optical component with desired properties.

Patterned by a laser, the thin film characteristics could not be matched by any other material.

The researchers put spin coated Graphene-oxide solution to a glass surface in order to produce the thin film.

With the help of a conventional laser system, they carved microstructures onto the Graphene-oxide film, by that tuning the nonlinear properties of the material.

In today's processes used for producing devices for telecommunications or optical devices, each component is manufactured individually and then engineers try to integrate them together.

Until now, no one could provide a film, on which everything can be fabricated with a laser and at the same time automatically integrate the components.

This process will allow faster speed and longer durability of the end products.

The biggest financial gain will come from reduced manufacturing cost.

The manufacturing processes used today in semiconductor plants require expensive clean rooms to produce defect-free photonic chips.

Now the fabrication and laser writing of this photonic material will become simple and cost will be dramatically reduced.

Clean rooms, costing several hundred million dollars, will become cheaper. (71)

Graphene-silicon photonic circuits enabling low-power optical information processing and frequency generation

The optical nonlinear behavior of Graphene is relevant for applications in optical interconnects and low power photonic integrated circuits. This is one of the major research fields for that wonder material.

In July 2014 a team from the engineering department of Columbia University explained to the press that they had been able to demonstrate and explain the strong nonlinear response from Graphene, which is the key component in this new hybrid device. [72]

Scientists placed a tiny layer of Graphene (just one-carbon-atom-thick) onto a passive device and thus transformed the originally passive device into an active one that could produce microwave photonic signals and converted wavelength of those microwaves in a way that made it useful for generating telecommunication wavelengths.

Prof. Chee Wei Wong "spoke about the 'magic' of Graphene's amazingly conductive properties. In his research, he exploited the optical nonlinearity behavior of Graphene. This is a property required for the digital on/off two-state switching and memory features.

By this method they produced new optical frequencies through mixing of two electromagnetic fields at low operating energies, allowing reduced energy per information bit. Underlying silicon structures served as a platform, showing that it is possible to use both materials, Graphene and silicon, in one device. This is similar to the work reported by IBM researchers previously.

Combining Graphene with silicon photonic crystal structures had long been considered impossible and many researchers believed that the "marriage" of these two systems would prohibit developing microcircuits based on Graphene. They were proved now to having espoused to a wrong idea.

This research paves the way for ultra-low power on-chip optical communications. This will lead to faster microcomputers, electronic devices and telecommunication system with stunning features.

Telecommunication companies as well as manufacturers of semiconductors soon might build first devices leveraging the new technology.

Making rubber bands with Graphene

The following might sound boring for high-technology computer nerds, but the research performed by a team from Surrey University and Trinity College in Dublin is extraordinarily compelling for material scientists and producers of consumer goods.

Scientists from the universities mentioned above turned simple rubber bands into body monitoring sensors by doping them with Graphene. (73)

The goal of the research was to create cheap strain sensors that work by measuring the oscillating change of flow of electricity through them as they move and stretch.

Measuring this change can be used to work out how much the band has stretched and, if the band is worn on the body, this can indicate how the user is moving in real-time.

This allows real-time monitoring of body-functions (e.g. In sports) but also could enable applications in other areas such as the automotive industry.

By measuring the temperature of the seat surface, automatic regulation of heat flow could be achieved. This is only one potential application of many.

Graphene will not only become a main component of electrical and optical devices; it will also change dramatically other simpler areas of technology. By that products will become cheaper and more reliable, as cheap 2D material Graphene will replace complex electronic devices.

This will also help the environment, as carbon, the material Graphene consist of, is bio-degradable and could be produced in zero-energy wasting a climate-neutral processes.

Another Graphene research related to supercapacitors

As mentioned before, supercapacitors are about to change the way we store energy and also how our cars will look and drive in the future.

Graphene has been seen by many engineers as a perfect material for enabling supercapacitors to combine the quick recharge and fast availability of energy of capacitors with the high storage capacity of conventional batteries.

Until August 2014 most research had not led to substantial progress in that area.

On that day a research team from the California NanoSystems Institute of University of California in Los Angeles published their findings about something they called a three-dimensional Graphene framework with porous structure, in short holey Graphene.

It is "holey" with an "e" *(not to muddle this up with "holy"...)* as the material that had been invented is porous and such full of tiny Nano-scale holes.

By that they could demonstrate electrodes that are capable of delivering a gravimetric capacitance of 298 Farads per gram (F/g) and a volumetric capacitance of 212 Farads per centimeter (F/cm).

If you're not a specialist in that specific engineering field, those numbers don't tell you very much.

It is only important to remember that this energy density is by far more than conventional lithium-ion batteries are capable of delivering.

However, there is much more research required because at this time the cost of Graphene based supercapacitors would by far exceed those of lithium-ion batteries.

There is some funny extension of this publication of UCLA's research team:

They engineered the "holey" device by means of a very sophisticated and complex engineering process.

Adopted to large scale manufacturing processes, this technique will become much cheaper over time and also be less expensive than today’s lithium-ion manufacturing process but still will remain expensive.

There might be some cheaper process be available in the future, if the findings described below turn out to become useable on production facilities.

What first sounded like a joke turned out to be serious scientific research with results that left the specialists in that area totally stunned.

Researchers in South Korea recently claimed that they found a competition to Graphene for energy storage.

The material that makes up used cigarette butts outperforms carbon, Graphene and carbon nanotubes in energy storage, they said. [74]

This material is a porous carbon material that has a pore structure that is perfect for energy storage in the electrodes of supercapacitors. This is similar to the "holey" Graphene found by UCLA's team.

What stays unclear with the results of that research team in South Korea is, why the claim to have found a material competing with Graphene.

As the paper, published on August 5, 2014, reveals, they found *"a porous carbon material that contains both the mesopores and micropores..."* and is mostly composed of carbon nanostructures.

Where is that different from Graphene embedded in a nanostructure grid?

The results, published in that paper are very intriguing, but I don't believe that they've found something else than porous Graphene and that their claims to have invented a much better material is simply an exaggeration.

Flexible Display made with Graphene

Graphene is more flexible than any other conventional ceramic alternatives used today in display technology and more transparent than metal films.

A team from the university of Cambridge Graphene Centre and Plastic Logic in UK presented a prototype of a Graphene based flexible display.

This display already has been engineered to a level of maturity that will allow usage in the production of consumer goods such as e-book readers or smart phones.

Photography courtesy of Plastic Logic

The device is an active matrix electrophoretic display, similar to the screens used in today's e-readers, except it is made of flexible plastic instead of glass.

Using Graphene in this process will allow to produce much cheaper and much more robust devices.

The glass of today's smart phones might be replaced eventually with material known to be nearly un-destroyable.

The research team believes that this breakthrough in display technology will enable a new generation of ultra-flexible and even foldable electronics.

Despite not being a core company in the field of Graphene, Plastic Logic has been added to the company list I'm observing for future performance.

Once they file an IPO, I'll definitely check them for their financial data.

More companies in unspecified research fields

Biogenic Reagents is not a typical company working on 2-D material development. This Minnesota based firm produces a special form of carbon products made from renewable sources.

Among those products one can find [46]:

- Activated carbon for mercury and emissions control
- Metallurgical carbon for iron and metals production
- Thermal carbon for energy generation

What makes this company attractive to investors is their impact on old-fashioned industries such as coal miners, water purification companies, steelworks or power plants.

The carbon produced by Biogenic Reagents is not much cheaper than competing products delivered by traditional sources – i.e. coal mines – but it can help reduce the environmental footprint left by coal producing companies and metal manufacturers.

Biogenic Reagents only recently started offering Graphene-carbon compounds, known as Ultra-Adsorptive Carbon ("UAC"), which are made from renewable biomass.

Biogenic Reagents wants to replace traditional activated carbon products for air and water purification with this new compound.

It should be very successful, as the new product is not only cheaper than traditional material but also more efficient.

As a startup company, Biogenic Reagents is led by CEO Jim Mennell, who has been quite successful in the past with renewable energy projects. He was able to raise millions in private capital, developed patent-pending processes, equipment and products, and built the largest advanced bio-carbon production facility in North America. [47]

Biogenic Reagents are not yet listed on the stock market. However, as soon as this firm becomes public, stocks should be watched carefully.

This is one of very few companies targeting a multi-billion dollar market with significant technology based on Graphene and carbon.

Third Millennium Materials, LLC

Located in Dayton, Ohio, Third Millennium Materials, LLC, under the leadership of CEO Louis A. Luedtke and chief of science officer Harry Couch, possesses enhanced material that might change many industrial and defense applications and processes.

They have named the new product "Covetics", or "carbon metal composite".

Initially, both founders worked as scientists at the National Composite Center in Kettering.

They received a U.S. patent for Covetics, and more than ten additional patent applications are pending.

Covetics enhances the properties of base metals, increasing its resistance against corrosion while giving metals unusual properties such as dramatically increased stiffness.

Covetics materials are a new class of carbon compounds made of metals and Nano-carbon structures that have bonded chemically.

Defense Advanced Research Projects Agency, or DARPA, as well as major industrial companies, started evaluating use cases for Covetics.

This material is already tested at U.S. naval bases. Private funding companies and Ohio's Third Frontier Advanced Materials Program have already invested 14 million US$ in that firm.

First, commercial applications of covetics will skip the military industry in favor of the luxury jewelry market.

This market can be addressed much faster than military projects, ensuring Third Millenium's financial stability by 2014.

Graphene Frontiers, a spinout from the University of Pennsylvania in the U.S., has been awarded $745 thousand from the National Science Foundation (NSF) to demonstrate and develop a roll-to-roll production of continuous Graphene film.

The company, founded in 2011, is using atmospheric-pressure chemical vapor deposition to produce Graphene on continuous tapes of copper foil passed through a growth region.

This eliminates the need for an expensive vacuum furnace and enables fabrication of Graphene films larger than the furnace size.

The funding will allow the firm to prove that sheets can be transferred from the metal catalyst to nearly any smooth surface without high-temperature steps or the use of harsh chemicals.

It's also worth noting that the transfer process preserves the original metal substrate for reuse, which both lowers production costs and reduces waste." (54)

Lockheed is definitely not a startup or small company.

Mainly known for its military planes, this company recently published another intriguing use case for Graphene. They made use of the thinness of Graphene.

Just one atomic layer thin, but stronger than steel or any other material, Lockheed engineers are capable of cutting tiny holes into ultra-thin sheets of carbon.

Just a few atoms in size, these cavities would only let desalinated water pass through. Since the sheet is ultra-strong at the same time, the amount of energy required for desalinating salt-water is decreasing dramatically to only a small percentage of what was needed in the past.

Territories consisting mainly of desert could profit from this. California or Australia, for example, might independently overcome low rain levels.

Combined with energy producing wind or solar farms, clean water may become an available resource everywhere for a low cost.

The material is already on the market. One could buy it under the name "Perforene".

"In addition to desalination, the Perforene membrane can be tailored to other applications, including capturing minerals, through the selection of the size of hole placed in the material to filter or capture a specific size particle of interest.

Lockheed Martin has also been developing processes that will allow the material to be produced at scale." [55]

Development areas at University and research institutes

This chapter focuses on current research activities which may lead to eventual spin-offs.

Looking to Korea, for instance, we find a tremendous amount of small research groups working hard and successfully on new products based on 2-D material.

Korea's KAIST Institute has three campuses, one in Seoul and two in Daejeon. Founded in the 1960s, it has developed into a leading research institute.

KAIST scientists used Graphene to create metals, hundreds of times stronger than commercial quality products. This research discovered a Graphene compound material 500 times stronger than copper or nickel.

"Graphene displays strengths 200 times greater than that of steel, is stretchable, and is flexible. The U.S. Army Armaments Research, Development and Engineering Center developed a Graphene-metal nanomaterial but failed to drastically improve the strength of the material.

To maximize the strength increased by the addition of Graphene, the KAIST research team created a layered structure of metal and Graphene. Using CVD (Chemical Vapor Deposition), the team grew a single layer of Graphene on a metal deposited substrate and then deposited another metal layer. They repeated this process to produce a metal-Graphene multilayer composite material, utilizing a single layer of Graphene. Micro-compression tests within Transmission Electronic Microscope and Molecular Dynamics simulations effectively showed the strength enhancing effects and the dislocation movement in the grain boundaries of Graphene on an atomic level." [50]

Another team at **KAIST** developed a new fabrication process for foldable Graphene circuits based on paper substrates. The new method uses a transfer printing process to prevent direct contact of the solvent and paper and to easily control the thickness of the deposited Graphene.

This makes it very simple to create foldable–not only flexible–electronics circuits. In addition, this makes it easier and more cost effective to create solar cells and electronic devices by simply printing cheap Graphene sheets.

The research team already demonstrated a working sample for an LED device.

One could imagine that eventually Graphene circuits will be printed on shirts.

When putting your (then foldable – see chapter 6) mobile phone or iPad into your pocket, it will be charged automatically by invisible Graphene-based conducting wires in that shirt's surface.

Michigan Technological University professor Yun Hang Hu has developed a new, inexpensive material that could replace the platinum in solar cells without degrading their efficiency: 3D Graphene.

Platinum, the key ingredient in certain solar-cells, is one of the most expensive metals on the planet. While only small amounts are needed, at $1,500 an ounce, the cost of the silvery metal is still significant.

"Hu and his team invented a novel approach to synthesize a unique 3D version with a honeycomb-like structure.

To do so, they combined lithium oxide with carbon monoxide in a chemical reaction that forms lithium carbonate (Li_2CO_3) and the honeycomb Graphene.

The Li_2CO_3 helps shape the Graphene sheets and isolates them from each other, preventing the formation of garden-variety graphite.

Furthermore, the Li2CO3 particles can be easily removed from 3D honeycomb-structured Graphene by an acid.

The researchers determined that the 3D honeycomb Graphene had excellent conductivity and high catalytic activity, raising the possibility that it could be used for energy storage and conversion.

So they replaced the platinum counter electrode in a dye-sensitized solar cell with one made of the 3D honeycomb

Graphene. Then they put the solar cell in the sunshine and measured its output." [51]

The cell with the 3D Graphene counter electrode converted nearly as much of the sun's energy into electricity as the conventional solar cell using costly platinum.

That development will change the way –and, of course, the price– of today's solar cells.

A similar invention was found at **the Fuda University in Shanghai in China**.

A team led by professor Gong discovered a practical way to reduce the amount of platinum needed for solar cells by a factor of 1.000.

This could reduce production costs for solar cells dramatically. [52]

Of **course MIT (Massachusetts Institute of Technology)** is participating in the race for Graphene applications, as well.

So far, it has been very difficult to develop grapheme-based semiconductor circuits. Texas Instruments and other companies had some success in building

Field Effect Transistors (FET) based on grapheme, but it was so tedious that many researchers stated that Graphene might not be capable of working as a semiconductor device.

Now scientists at MIT have managed to develop a semiconductor device based on Graphene by simply doping it with chlorine.

The material could be used to make grapheme-integrated circuits on a chip and RF transistors and more.

Doping the Graphene with chlorine also reduces sheet resistance, making it suitable for use in electronic circuit interconnects.

The team, led by M. Dresselhaus and Tomas Palacios, plans to dope suspended samples of Graphene with chlorine–to access both sides of a sample–and open up an even bigger electronic band gap. [53]

Monash University

One of the biggest challenges for renewable energy to become successful and completely replace fossil fuel is the problem of storing energy produced by wind farms, solar cells or other devices.

At Monash University, a team of scientists led by Professor Dan Li of the Department of Materials Engineering developed a completely new process to develop Graphene-based supercapacitors.

These capacitors could become the “missing link” between energy-producing wind farms, solar fields and industries which rely on fossil fuels.

On top of this new invention, they created their material using a method similar to that used in traditional papermaking, meaning the process could be easily and cost-effectively scaled up for industrial use.

Researchers at the **University of California, Berkeley**, investigated a completely different market. They created the first ever Graphene audio speaker: an earphone.

Using Graphene’s superior physical and electrical properties, the earphone showed a frequency response comparable to or better than a pair of commercial Sennheiser earphones.

It could surpass **Bose, Sennheiser, or Sony** if they are not capable of making use of this invention.

Prospects and vision for future developments in 3D-printing and Internet of Things (IoT)

3D printing

3D printing often is considered to be the start of the next industrial revolution, as is 2-D nanotechnology or the Internet of Things (IoT).

In my opinion, it'll turn out that the combination of these three cutting edge technologies will become the foundation of the next industrial revolution, not one single of those mentioned technologies.

3D printing has the potential to bring us to a new reality of customizable, one-off production. Adding 2-D nanotechnology to that, this definitely will become true.

Industries like aerospace, automotive, defense, telecommunication, and medical are already taking advantage of 3D printing.

Market gurus expect that 3D printing will be as significant for modern manufacturing as the invention of the assembly line by Henry Ford. That process triggered the second industrial revolution.

By combining „fused deposition modeling" with 2-D nanotechnology in a 3-D printing process one eventually will be capable of producing devices that are stronger, more flexible and customizable than anything thought about before.

Integration energy storage and sensor technology in the same production process will make IoT devices and gadgets possible, we've never considered they could ever exist.

> *„Fused deposition modeling (FDM) is an additive manufacturing technology commonly used for modeling, prototyping, and production applications.*
>
> *The term fused deposition modeling and its abbreviation to FDM are trademarked by Stratasys Inc.*
>
> *The exactly equivalent term, fused filament fabrication (FFF), was coined by the members of the RepRap project to give a phrase that would be legally unconstrained in its use. It is also sometimes called Plastic Jet Printing (PJP).*
>
> *„FDM works on an "additive" principle by laying down material in layers; a plastic filament or metal wire is unwound from a coil and supplies material to produce a part.*
>
> *„The technology was developed by S. Scott Crump in the late 1980s and was commercialized in 1990.*
>
> *(Source: Wikipedia (75))*

In Fused Deposition Modeling (FDM) or Plastic Jet Printing plastic polymers are heated and then deposited layer-by-layer through an extruder to create a physical object.

> *"It is also the most widely used method of 3D printing and the method in which our current technology is based on.*
>
> *FDM printers have been used in a variety of disruptive ways, including: 3D printed end use parts for NASA's Mars Rover, sporting goods, and medical devices.*
>
> *FDM 3D printing has even been used to create affordable prosthetics and an exoskeleton to assist the disabled.*
>
> *There are certain current limitations of the 3D process, in particular the inability to combine several functional materials within a single print.*
>
> *Different materials (such as metals) require different parameters to be successfully printed, including extreme temperatures.*
>
> *Also, 3D printed products are not as inherently strong as traditionally manufactured parts.*
>
> *By compiling intricate 3D-printed components such as sensors, batteries, light and power sources, and electronic circuits, complex smart objects can be created.*

With the inclusion of integrated GPS and wireless chips, the devices can even be connected to the Internet and controlled remotely.

A few examples of future applications include security devices that are undetectable by criminals and compact autonomic sensors that can detect oil spills or toxic gases and notify authorities.

Such technology will even enable the development of cutting-edge, cost-effective wearable devices, from electronic gadgets to health monitors. "

(Source: graphene3Dlab [60])

All this will change when 2-D materials, especially Graphene, will be available for 3-D printing technologies.

Then the core material won't be weak and unstable polymers but the strongest and most perfect material known today.

As of today, there are two companies claiming to be ready for fueling this compelling market. One is previously mentioned graphene3Dlab [60], the other one an Australian startup called 3D Group [76].

Both companies, graphene3Dlab as well as 3D Group, should be carefully observed.

It seems that graphene3Dlab are way out in front of 3D Group in terms of intellectual property.

One only can be sure about investment opportunities in both companies when production has started.

Competing products to Graphene

In the early 1980s, I worked as a student in a team at the University of Aachen (Aix la Chapelle is the French name).

Our group consisted of physicists and engineers who had a challenging mission.

We wanted to be the first team in Europe to grow II-VI semiconductors by means of a new method.

The roman numbers II and VI (2 and 6) refer to the position of a chemical element in the periodic table of elements.

All these elements have a number referring to its position in that table.

Carbon and Silicon, for instance, belong to group IV, the Latin number 4.

Along with Germanium these are the well-known semiconductors.

> *(The first semiconductor element which was detected is Selenium. Willoughby Smith, a British engineer found the photoelectric effect in that stuff. Selenium belongs to group six. In order to understand why group four elements as well as*

those from group six show semiconductor properties, one has to apply principles of quantum physics. Here we are again, as Graphene properties and those of any 2-D material could only be understood and predicted by using complicated mathematical computing methods based on quantum physics.)

By mixing or alloying base elements of the third and the fifth group, such as Gallium and the poisonous Arsenide, one also could get semiconductors with similar or even better electronic properties than those of group four (i.e. Silicon). That was a well-known fact for many years.

Also elements from the second and sixth group, such as Indium and Telluride, were known to behave like those from the fourths group, i.e. showing semiconductor behavior.

Many scientists believed that those II-VI materials, as they were denoted, were superior to silicon and gallium-arsenide.

Especially for optical switches they should provide the features necessary to enable large fiber-optic networks.

This turned out to be true and today most switches are based on II-VI or III-V compounds or a mix of those.

Most of these scientists also were convinced that these compounds and alloys eventually would replace silicon in semiconductor devices.

That turned out to be wrong.

Whenever any expert in the last decades had announced to have found some effect which would make silicon obsolete – and there were many – silicon stroke back and passed its competitors.

For decades this was like the race between the hedgehog and the tortoise.

Whenever gallium-arsenide or one of her friends from the II-VI section arrived at the finish, silicon was already there.

In 2004 the research community gave up and bowed their thanks to silicon.

That material had outperformed any prediction and withstood all mathematical theories denying its superiority.

Since the advent of Graphene many scientists believe that the "twilight of God silicon" has appeared.

Graphene based technology should be able to do what Gallium-Arsenide or other compound semiconductors weren't capable of achieving.

Let's come back to the work I did in the semiconductor lab in Aachen.

I was part of the race game between silicon and II-VI and III-V for many years.

In 1982 I entered the team, which was led by Holger Jürgensen and Meino Heyen.

Holger is a charismatic scientist who planted the idea in young students of his team that alloys such as Indium-Telluride and others would be the future of optoelectronics and the computer industry.

Holger himself, along with Meino Heyen, was not particularly interested in finding competing products to silicon.

They wanted to create optoelectronic devices. Period.

But other scientists believed to be capable of replacing silicon by other material.

Our team was part of that race. Most of us probably understood this many years later to the full extent only.

Meino had invented the first MOCVD system in the university and as I remember, it was one of the first MOCVD systems in the world.

MOCVD stands for Metal-Organic Chemical Vapor Deposition.

This process uses poisonous gases such as phospine, arsine or trimethyl-indium for creating optical semiconductors.

If you want to learn about that I suggest searching the Internet for MOCVD. There are some great articles about it.

One morning in that year I came to the building where we had our laboratory, just to find it blocked by police and fire brigades.

The evening before somebody had forgotten to close a valve on one of the reactors and chemicals reacted with the oxygen of the air, causing a fire.

That blaze destroyed all experiments and a group of aspiring scientists lost all they'd had over night.

It turned out to be less harmful as we'd initially believed.

The laboratory could be reconstructed within a few months and one year later Holger and Meino could found their company.

They called it Aixtron.

This is a blend of the name of the town of the university (Aix La Chapelle) and the word Electronics.

After the fire I'd left the team on amicable terms, I had to work in order to finance my study and couldn't afford to extend my study any longer, since a wife and three kids expected more from me than scientific excellence.

So I started working on a silicon based project in another university.

That was how I entered the silicon sphere.

Aixtron became a successful producer of CVD systems and is now one of the market leaders in that field.

They also should be capable of leading the pack in the Graphene core market

Some years later I finished my doctoral thesis about the interaction between silicon and boron-nitride, where I found the hydrogen storage capabilities of that system.

During that work I faced some severe problems. Ultra-thin Nano filaments of single layer carbon atoms in the bulk silicon sometimes drove me nuts. I guess I'd entered into something that was understood only 20 years later and then named Graphene.

Shortly after that I started my career in the industry in a silicon semiconductor fab of Texas-Instruments.

To cut a long story short, why do I tell this story?

Today we see Graphene emerging as a superstar for future applications. It'll change our lives dramatically.

As described before, this material is a super-product in all realms of material technology.

There are plenty of other materials known to have 2-D properties.

Among them are Boron-Nitride, Tantalum-Sulfide, Molybdenum-Sulfide, Niobium-Selenide, Phosphorene, Stanene (2D tin) and many others.

So far most of these compounds, except Molybdenum-Sulfide, have shown to be more difficult to be handled and providing worse capabilities than Graphene.

These alloys could only play a role in serving Graphene.

That might change. The tortoise seems to have returned after a short break.

A scientist with the name Guy Le Lay from the university of Aix-Marseille in 2012 found a version of silicon that he called silicene. He named it that way because he wanted it to sound similar to Graphene.

Silicene is the 2-D version of silicon and as its competitor (or friend, who knows) it's a one-atom thin layer of silicon atoms. Unfortunately, its properties are worse than those of Graphene.

It's chemically unstable, it likes to crinkle and it is difficult to produce.

It seems that silicon finally lost the race. A new material is there which conquered silicon.

I don't believe that.

Both Graphene and silicene show excellent features in mathematical models.

Together they could make this disruptive technology much more reliable. The combination of both materials could help overcome many of the obstacles seen today in Graphene based applications.

Silicene layers between Graphene layers could help tune the quantum band gap between both systems, and they could help in reducing the complexity of manufacturing Graphene devices.

The technology to achieve that is already available. It's called CVD.

Earlier in this book I mentioned that Graphene could also become a simple storage system for hydrogen. As I said, I've not found any scientific paper supporting this idea.

When I wrote this, I had in mind what problems I'd faced during my research in the 1980's with the system silicon-boron-nitride and the unwanted carbon filaments in it.

Those had a diameter of 2-3 atomic layers or less.

This was where I found hydrogen (by infrared spectroscopy).

But I had no hydrogen in my laboratory and before alloying the devices there was no hydrogen to be detected.

I would not be surprised if carbon and silicon layers had attached to each other and formed a quantum band gap which was a perfect fit for small atoms to tunnel through.

And that small molecule is hydrogen, simply extracted from the water I used to clean the small silicon pieces with.

For those not familiar with the scientific "gibberish": the system silicene-Graphene simply could act like a sponge for hydrogen, which is not controllable under normal circumstances, i.e. room temperature and normal pressure.

Could it be that at the end the hedgehog and the tortoise are concluding peace and become friends?

I would like this idea. But it well may be that the story continues on a different path.

There is one 2-D product showing similar and in some terms even better 2-D properties than Graphene.

Researchers today investigate another 2D material—molybdenum sulfide (MoS_2) —which has similar characteristics as Graphene but offers something Graphene doesn't: a wide energy band gap.

This enables transistors and circuits to be built from it directly. [42]

For some electronic devices molybdenum sulfide could soon replace Graphene.

How would that affect the stocks of companies that had invested in Graphene?

From all what we already see in the technological area, both Graphene and molybdenum sulfide could be used in the same way.

So any company capable of using Graphene might as well be able to us MoS_2.

Also, boron-nitride and still not detected groups of 2-D material will be used eventually.

There will be applications for all 2-D material, where all find its respective niche. All of the 2-D stuff is ready for use and it will be applied in all areas mentioned in this book.

Our job as investors is to identify those applications and let the scientists find out what material suits their needs best.

Graphene research is by far ahead of any other 2-D material.

Nevertheless, it might be replaced by another material one day.

But this would only affect the Graphite mining companies.

Application vendors should be able to switch between any of the mentioned chemicals.

What is the risk?

There are many voices saying that Graphene might never become as successful as silicon in the past.

This opinion I hear frequently when discussing with business people, journalists and also reputable scientist.

One argument is that it will take decades for Graphene to evolve at that large-scale silicon has occupied today, because silicon itself took decades to evolve and the related technologies had not been understood at that time.

Graphene transistors are compared to the primitive first silicon transistor Mr. Bardeen, Mr. Shockley and Mr. Brattain developed at the Bell Labs in 1947.

At that time nobody had the vision of integrating these devices into a microchip and it took a full generation until the first microprocessors showed up in laboratories.

Today, we have the vision.

Silicon technology has paved the way to the Internet and beyond.

When I look back into the year 1990, many journalists and tech gurus claimed that the Internet never would be more than a niche application.

During this book I've shown how many companies are investing billions in 2-D technology and that many applications are already on the horizon.

Applications, which, some people said 5 years ago, would never be possible before 2030 – if ever – were demonstrated by companies such as IBM already in 2011 and 2012.

There is one truth in the skepticism of the people mentioned above.

In 20 years from now we'll see technologies we'd never thought about.

Our forefathers could imagine computer systems already in the early 20Th century, but nobody had any vision about the personal computer, the Internet, optical switches or gadgets like an iPad.

All these nice things we have today evolved through a sudden inspiration of brilliant young engineers and scientists.

So it will be in the future.

The breathtaking speed of scientific developments will let new small companies thrive in new areas of technology, biology, science, engineering and so on.

We'll see self-organizing 3-dimensional computers, based on Graphene.

DNA computers will be integrated with inorganic devices and again with Graphene.

Optical systems will empower blind people to see again.

A completely new biological industry will evolve with these new technologies.

And so on.

Ray Kurzweil, the famous information technology guru, not only predicted many of these things, but also showed that this kind of exponential growth is powered by changes in paradigms. [19]

By 2020 we'll see the end of the Moore's law for today's silicon technology.

By that time Graphene will enable another exponential law.

And may be that silicen or another 2-D material will trigger the next step.

By that time, many new companies will have emerged with great ideas and products.

Those are the ones to check out for investments.

The winners in terms of financial prosperity will be those people who invested into the stocks of the right companies at the beginning.

Summary:

In this book (which will be replenished regularly) I've tried to relay a few of the breathtaking developments taking place NOW. These inventions are no longer science fiction and, very soon, consumers will see great new products and widgets based on 2-D material, mainly Graphene.

Compile the ideas above into a common picture:

- Electronic devices will become foldable
- Integrated circuits will be printed on paper
- Solar cells will become as cheap as a sheet of paper
- ...

When I started writing this book, I believed that most of the inventions would never take place before the next decade.

However, the ongoing successes in research have changed my opinion.

Graphene and 2-D based material paradigm shifts are about to happen within the next five or so years.

Some of these new developments will hit certain conservative industries like an earthquake.

I don't believe that fossil fuel producing companies will get out of this scenario unscathed.

Do you remember the story about the railway industry in the twentieth century, which is told to every sophomore in business school?

Nearly all of those companies died when planes became the leading devices for mass transportation.

Why was that?

The reason is simple: The management of those old-fashioned firms believed that they were in the railway business.

As soon as they realized that their business was not railways, but instead delivering goods from point A to B, it was too late.

Only a few of those old-fashioned companies survived.

A similar scenario will hit the fossil fuel industry, renewable energy, steel makers, companies in the market for integrated circuits, and consumer goods, to name a few.

Studying the management of firms in these markets will soon separate the wheat from the chaff.

Companies who utilize the new inventions and adapt their core processes to make use of them will enter into a great future.

The rest, simply, will vanish.

The end of this book, but not of the story

One last plea:

On the next page you will be given the opportunity to rate the book and share your thoughts on Twitter, Facebook or Google+

If you believe the book is worth sharing, would you please take a few seconds to let your friends know about it?

If it turns out that they'll enjoy it, they'll be forever grateful to you.

As I will.

All the best,

Dr. Elmar P. Selbach

Addendum A: Websites

Websites of companies mentioned in this book:

Grafoid: http://www.grafoid.com

Focus Graphite: http://www.focusgraphite.com/

Lomiko Metals: http://www.lomiko.com/

Northern Graphite: http://www.northerngraphite.com/

Aixtron: http://www.aixtron.com/en/home/

Vorbeck: http://vorbeck.com/

Graphene Nano Chem: http://www.graphenenanochem.com/

XG Sciences: http://xgsciences.com/

GRAnPH Nanotech: http://www.granphnanotech.com/

Crayo Nano: http://crayonano.com/technology/

Angstron: http://www.angstronmaterials.com/

AMO GmbH: http://www.amo.de/

Nantero Inc. : http://www.nantero.com/

+ISO: http://piuiso.com/

AMS Technologies: http://www.amstechnologies.com/

AsTEK: http://newweb.astekcorp.com/

C3Nano: http://www.c3nano.com/

Inteltronic: http://www.inteltronicinc.com/

ChemPacific: http://www.chempacific.com/

CreaPhys: http://www.creaphys.com/

Dresden Microdisplays: http://microdisplay.biz/

AU Optronics: http://auo.com/?sn=101&lang=en-US

Advantech: http://www.advantech.com/default.aspx

Cambridge Display: http://www.cdtltd.co.uk/

CSEM Switzerland: http://www.csem.ch/site/

COMEDD Germany: http://www.comedd.fraunhofer.de/

Yunnan North OLiGHTEK :
http://www.olightek.com/english/index.jsp

Vorbeck: http://vorbeck.com/

Dupont: http://www2.dupont.com/home/en-us/index.html

BASF: http://www.basf.com/group/corporate/en/

Bayer: http://www.bayer.com/

Sony: see for instance:
http://techon.nikkeibp.co.jp/english/NEWS_EN/20120828/236214/

Samsung (with its affiliate CheilIndustries):
http://www.samsungchemical.com/jsp/eng/pr_center/sm_pr_showroom.jsp

IBM: http://www-03.ibm.com/press/us/en/pressrelease/36135.wss

Henkel: http://www.henkel.com/index.htm

Kimberly-Clark: http://www.kimberly-clark.com/

Trumpf: http://www.trumpf.com/

AMS Technologies: http://www.amstechnologies.com/

Boston Applied Technologies: http://www.bostonati.com/

VEECO: http://www.veeco.com/

Apple: http://www.patentlyapple.com/patently-apple/2013/03/future-iphones-may-take-advantage-of-graphene-for-integrated-solar-cell-panels.html

Nokia: http://research.nokia.com/page/11906

Texas Instruments:
http://www.ti.com/corp/docs/innovation/invention.html

EADS: http://www.airbus.com/newsevents/news-events-single/detail/european-union-selects-the-graphene-project-for-research-funding-opening-prospects-in-aviation-and/

Qiagen: www.qiagen.com

Amgen: www.amgen.com

Gilead Sciences: www.gilead.com

ARIAD Pharma: www.ariad.com

Seattle Genetics: www.seagen.com

Biogen: www.biogenidec.com

Merck KGaA: www.merckgroup.com

Baxter: www.baxter.com

Actelion: www.actelion.com

Regeneron : www.regeneron.com

Forest laboratories: www.frx.com

Celgene: www.celgene.com

Elan: www.elan.com

ABB Robotics: www.abb.com/

Adept Technology: http://www.adept.com/

AeroVironment: http://www.avinc.com/

Aethon: http://www.aethon.com/

Aldebaran Robotics: http://www.aldebaran-robotics.com/en/

Astrobotic Technology: http://www.astrobotic.com/

Barrett Technology: http://www.barrett.com/robot/index.htm

Bluefin Robotics: http://www.bluefinrobotics.com/

Bossa Nova Robotics: http://www.bnrobotics.com/

Boston Dynamics: http://www.bostondynamics.com/

Clearpath Robotics: http://www.clearpathrobotics.com/

Cyberdnye: http://www.cyberdyne.jp/english/

DENSO Robotics: http://www.densorobotics.com/world/

EKSO Bionics: http://www.eksobionics.com/

Energid Technologies: http://www.energid.com/

EPSON Robots: http://robots.epson.com/

FANUC Robotics: http://www.fanucrobotics.com/

FESTO: http://www.festo.com/net/startpage/

FMC Schilling Robotics:
http://www.fmctechnologies.com/Schillingrobotics.aspx

Hansen Medical: http://www.hansenmedical.com/

General Electric (medical robots): http://www.ge.com/

Honeybee Robotics: http://www.honeybeerobotics.com/

Hyundai Heavy Industries: http://english.hhi.co.kr/main/

InTouch Health: http://www.intouchhealth.com/

Intuitive Surgical: http://www.intuitivesurgical.com/

iRobot Corp.: http://www.irobot.com/global/de/home.aspx

John Deere Mobile Robotics:
http://www.deere.com/wps/dcom/globalhome/deerecom/global_home.page

Kiva Systems: http://www.kivasystems.com/

KUKA Robotics: http://www.kuka.com/

Lely: http://www.lely.com/en/home

Liquid Robotics: http://liquidr.com/

MAKO Surgical: http://www.makosurgical.com/

MathWorks: http://www.mathworks.com/

Medrobotics: http://www.medrobotics.com/

Northrop Grumman:
http://www.northropgrumman.com/Pages/default.aspx

RedZone Robotics: http://redzone.com/

Rethink Robotics: http://www.rethinkrobotics.com/

Robosoft: http://www.robosoft.com/

Robotiq: http://robotiq.com/en/

Schunk:
http://www.schunk.com/index_select_your_country2.html?r=1

Seegrid: http://www.seegrid.com/

Stäubli: http://www.staubli.com/

Swisslog: http://www.swisslog.com/de/?redirected=true

Universal Robots: http://www.universal-robots.com/

Vecna Technologies : http://www.vecna.com/

VGo Communications: http://www.vgocom.com/

Willow Garage: https://www.willowgarage.com/

Yaskawa Motoman Robotics: http://www.motoman.com/

Addendum B: Reference & Citations

(1) "The Nobel Prize in Physics 2010". Nobelprize.org. 29 Apr 2013 http://www.nobelprize.org/nobel_prizes/physics/laureates/2010/

(2) "The 2010 Nobel Prize in Physics - Press Release". Nobelprize.org. 29 Apr 2013 http://www.nobelprize.org/nobel_prizes/physics/laureates/2010/press.html

(3) Tbd.

(4) Graphene Flagship Proposal Report – European Commission 2012

(5)https://www.youtube.com/watch?v=lX-gTobCJHs

(6) http://conversations.nokia.com/2011/06/14/graphene-nokias-future-super-material/

(7)-(11) were taken from the website http://www.rinkworks.com/said/ but also could be found in http://en.wikiquote.org/

(12) Taken from http://en.wikiquote.org/ (as quoted in *The Harvest of a Quiet Eye : A Selection of Scientific Quotations* (1977), p. 56)

(13) In the literature about Graphene I've found some indicators for using Graphene as a hydrogen storage device. (e.g. http://arxiv.org/ftp/arxiv/papers/1207/1207.5703.pdf)

My own scientific work in the 1980th showed that material, which at that time came close to 2-D properties, was able of storing hydrogen at room temperature and dissolving it slowly into air if heated beyond 200° Celsius. The material I used was Boron-Nitride film on silicon.

(14) http://www.graphenetracker.com/invest/

(15) http://www.eetimes.com/electronics-products/electronic-product-releases/analog-products/4203211

(16) http://www.eetimes.com/electronics-news/4371446/IBM-demos-terahertz-graphene-photonics

(17) http://www.nist.gov/pml/div683/conference/

(18) Cheianov, Fal'ko, Altshuler, Science (2007)

(19) http://www.kurzweilai.net See also: Ray Kurzweil: A university for the coming singularity. Speech at TED conference. www.ted.com

(20) http://www.engadget.com/2012/05/18/samsung-graphene-breakthrough/

(21) http://www.londonstockexchange.com/exchange/prices-and-markets/stocks/new-and-recent-issues/new-recent-issue-details-description.html?issueId=8835

(22) http://www.cvdequipment.com/

(23) http://crayonano.com/markets/

(24) http://www.angstronmaterials.com/

(25) http://www.amo.de/4.0.html?&L=1

(26) http://www.graphenea.com/

(27) http://nanoridge.com/

(28) http://www.swentnano.com/

(29) http://2-dtech.com/

(30) http://americangraphitetechnologies.com/

(31) http://bluestonegt.com/

(32) http://www.clbattery.com/

(33) http://www.durhamgraphene.com/

(34) http://vulvox.tripod.com/

(35) http://xolve.com/

(36) http://sinodesystems.com/

(37) http://www.src.org/program/nri/

(38) http://investing.businessweek.com/research/stocks/private/snapshot.asp?privcapId=1083480

(39) www.elsevier.com/locate/biomaterials

(40) http://www.genomeweb.com/sequencing/qa-henk-postma-graphene-nanogaps-electronic-single-molecule-sequencing

(41) http://www.extremetech.com/computing/153046-mit-and-harvard-engineers-create-graphene-electronics-with-dna-based-lithography

(42) http://www.electroiq.com/articles/sst/2012/september/molybdenum-sulfide-the-new-graphene.html

(43) http://e2e.ti.com/blogs_/b/on_the_fringe_with_gene_frantz/archive/2013/03/01/graphene-s-potential-to-extend-moore-s-law.aspx

(44) http://www.src.org/program/starnet/

(45) http://www.ft.com/intl/cms/s/0/c1f4ad36-45d8-11dc-b359-0000779fd2ac.html#axzz2UUhewYYm

(46) http://www.biogenicreagents.com/about/companyinformation/

(47) http://www.biogenicreagents.com/news-and-events/biogenic-reagents-opens-largest-advanced-biocarbon-production-facility-in-north-america/

(48) http://www.azom.com/article.aspx?ArticleID=8151

(49) http://www.garmortech.com/

(50) https://www.kaist.ac.kr/english/01_about/06_news_01.php?req_P=bv&req_BIDX=10&req_BNM=ed_news&pt=17&req_VI=4443

(51) http://www.ncbi.nlm.nih.gov/pubmed/23897636

(52) http://www.ncbi.nlm.nih.gov/pubmed/21909512

(53) http://www-mtl.mit.edu/wpmu/tpalacios/

(54) http://tmrplus.iop.org/

(55) http://www.lockheedmartin.com/us/news/press-releases/2013/march/lockheed-martin-achieves-patent-for-perforene-filtration-solutio.html

(56) http://www.appliedgraphenematerials.com/

(57) http://rareearthmetalssite.com/finance-watchdog-issues-warning-over-investing-in-miracle-material-graphene/

(58) http://www.fca.org.uk/news/graphene

(59) http://www.cortalconsors.com

(60) http://www.graphene3dlab.com/s/company.asp#about-us

(61) http://www.nanalyze.com/

(62) http://xgsciences.com/graphene-manufacturer/history/

(63) http://www.planartech.com/

(64) http://2-dtech.com/

(65) http://grapheneindustries.com/

(66) http://crayonano.com/about/

(67) http://investorintel.com/graphite-graphene-intel/chinas-graphene-industry-starts-take-2014/)

(68) http://www-03.ibm.com/press/uk/en/pressrelease/43928.wss

(69) http://www.forbes.com/sites/tomkonrad/2013/09/18/graphene-stock-investing-what-the-pros-think/

(70) http://phys.org/news/2014-08-tiny-graphene-future-quantum-memory.html

(71) http://phys.org/news/2014-05-graphene-photonics-breakthrough-fast-speed-low-cost.html

(72) http://www.columbia.edu/

(73) http://www.theengineer.co.uk/more-sectors/electronics/news/graphene-turns-rubber-bands-into-stretchy-body-sensors/1019136.article#comments_form

(74) http://iopscience.iop.org/0957-4484/25/34/345601/

(75) http://en.wikipedia.org/wiki/Fused_deposition_modeling

(76) http://www.3dg.com.au/

Pictures:

P-1: *© jamenpercy - Fotolia.com*

P-2: © B. Wylezich - Fotolia.com

P-3: © arsdigital - Fotolia.com

P-4: © vvoe - Fotolia.com

P-5: © martin_matthews - Fotolia.com

P-6: © Lightboxx - Fotolia.com

P-7: © lassedesignen - Fotolia.com

P-8, P-9: © Dr. Elmar P. Selbach

P-10: © frank peters - Fotolia.com

P-11: © Microstock Man - Fotolia.com

P-12: © frog - Fotolia.com

P-13: © Andre Montaud - Fotolia.com

14356184R00123

Printed in Great Britain
by Amazon.co.uk, Ltd.,
Marston Gate.